TM 9-331

155-MM HOWITZER M1 AND 155-MM HOWITZER CARRIAGE M1

TECHNICAL MANUAL

MARCH 1, 1943

by WAR DEPARTMENT

DISCLAIMER:

This manual is sold for historic research purposes only, as an entertainment. It contains obsolete information and is not intended to be used as part of an actual operation or maintenance training program. No book can substitute for proper training by an authorized instructor.

©2013 Periscope Film LLC
All Rights Reserved
ISBN#978-1-937684-37-2
www.PeriscopeFilm.com

TECHNICAL MANUAL
No. 9-331

WAR DEPARTMENT
Washington, March 1, 1943

155-MM HOWITZER M1 AND 155-MM HOWITZER CARRIAGE M1

Prepared under the direction of the
Chief of Ordnance

CONTENTS

	Paragraphs	Pages
CHAPTER 1. Introduction	1–4	2–8
CHAPTER 2. Howitzer and carriage	5–69	9–116
SECTION I. Description and functioning of howitzer	5–10	9–28
II. Description and functioning of carriage	11–25	29–54
III. Operation	26–39	55–63
IV. Malfunctions and corrections	40–43	64–71
V. Lubrication	44–46	72–77
VI. Care and preservation	47–57	78–93
VII. Inspection and adjustment	58–62	94–99
VIII. Disassembly and assembly	63–69	100–116
CHAPTER 3. Sighting and fire-control equipment	70–85	117–159
SECTION I. Sighting equipment	70–78	117–134
II. Fire-control equipment	79–85	135–159
CHAPTER 4. Ammunition	86–100	160–178
CHAPTER 5. Organization spare parts and accessories	101–102	179–180
CHAPTER 6. Storage and shipment	103–106	181–185
CHAPTER 7. Operation under unusual conditions	107–111	186–188
CHAPTER 8. Materiel affected by chemicals	112–114	189–191
CHAPTER 9. References	115–117	192–193
INDEX		194–203

1

TM 9-331
1-2

155-MM HOWITZER M1 AND 155-MM HOWITZER CARRIAGE M1

CHAPTER 1

INTRODUCTION

	Paragraph
Scope	1
Characteristics	2
Differences among models	3
Data	4

1. **SCOPE.**

    a. This technical manual is intended to serve temporarily (pending the publication of a more complete revision) to give information and guidance to personnel of the using arms charged with the operation, maintenance, and minor repair of this materiel. Some deviations from the standard nomenclature occur herein since the compilation of the standard nomenclature list had not been completed at the time the manual was prepared.

    b. In addition to brief descriptions of the 155-mm Howitzer M1 and the 155-mm Howitzer Carriage M1, this manual contains technical information required for the identification, operation, inspection, and care of the materiel.

    c. This manual contains descriptions of the authorized ammunition for this weapon as well as information required for the identification, operation, and care of the sighting and fire-control equipment authorized for use with this weapon.

    d. Certain disassembly, assembly, adjustment, and repairs of the materiel that may be handled by the using arms personnel are prescribed in this manual. They will be undertaken only under the supervision of an officer or the battery or artillery mechanic.

    e. In all cases where the nature of the repair, modification, or adjustment is beyond the scope of the facilities of the unit, the responsible ordnance service should be informed in order that trained personnel, with suitable tools and equipment, may be provided, or proper instructions issued for the performance of the work.

2. **CHARACTERISTICS.**

    a. *Howitzer.* The 155-mm Howitzer M1 is a short-barreled weapon using separate-loading ammunition. It throws projectiles of approximately 95 pounds at a muzzle velocity of 1,850 feet per second to a maximum range of approximately 16,500 yards. The rate of fire for

2

INTRODUCTION

Figure 1 — 155-mm Howitzer M1 and 155-mm Howitzer Carriage M1 — Firing Position — Maximum Elevation

TM 9-331
2-4

155-MM HOWITZER M1 AND 155-MM HOWITZER CARRIAGE M1

rapid bursts is 3 rounds per howitzer per minute; for prolonged firing, 1 round per howitzer per minute. The howitzer is equipped with a manually operated breech mechanism and a percussion type firing mechanism.

   b. *Carriage.*

   (1) The 155-mm Howitzer Carriage M1 is of the single-axle, 2-wheel, split-trail type. In traveling position, the trails are locked together by a toggle type clamping mechanism and are limbered directly to the prime mover. The wheels are equipped with large pneumatic tires for high-speed transport. When limbered to a prime mover, the materiel can be drawn at speeds up to 30 miles per hour on improved roads. The relatively moderate weight of the materiel adds to its maneuverability and its ease of transport.

   (2) The howitzer is fired from 3-point suspension, with the trails spread and the carriage resting upon an integral firing jack, the wheels being clear of the ground. In firing position, the howitzer has a range movement in elevation of 1,156 mils (65 degrees), and a traverse range of 942 mils (53 degrees), or 471 mils (26½ degrees), to the right and left of mid-position.

   (3) The howitzer is equipped with Recoil Mechanism M6 of the variable recoil, hydropneumatic type. The length of recoil varies from 60 inches at 0- to 25-degree elevation to 40 inches at 40- to 65-degree elevation. The carriage is equipped with electric power brakes with an emergency brake application mechanism controlled by a safety switch. Hand brakes are provided for parking. Armor plate shields are furnished for the protection of the operating personnel.

3. DIFFERENCES AMONG MODELS.

   a. There are no differences among models of howitzers and carriages which affect service by the using troops.

   b. While the initial production of carriages is being equipped with electric power brakes, consideration is being given to equipping carriages produced at some future date with air brakes.

4. DATA.

   a. **General Data Pertaining to the 155-mm Howitzer M1.**

Weight of 155-mm Howitzer M1, complete	3,825 lb
Caliber	155-mm or 6.102 in.
Length of bore (in calibers)	23
Length (muzzle to rear face of breech ring)	149.2 in.
Type of breechblock	Stepped-thread, interrupted screw
Chamber capacity	725 cu in.
Muzzle velocity	1,850 ft per sec

TM 9-331
4

INTRODUCTION

Figure 2 — 155-mm Howitzer M1 and 155-mm Howitzer Carriage M1 — Firing Position — Three-quarter Rear View

TM 9-331
4

155-MM HOWITZER M1 AND 155-MM HOWITZER CARRIAGE M1

Figure 3 — 155-mm Howitzer M1 and 155-mm Howitzer Carriage M1 — Limbered to Prime Mover

INTRODUCTION

Muzzle energy	2,260 ft-ton
Maximum powder pressure	32,000 lb per sq in.
Rifling:	
Length	113.10 in.
Number of grooves	48
Twist	Uniform, right-hand, one turn in 25 calibers
Weight of projectile	95 lb
Weight of powder charge	5.94 to 13.86 lb
Travel of projectile in barrel	120.675 in.
Maximum range with supercharge	16,500 yd
Rate of fire:	
Rapid bursts	3 rounds per howitzer per minute
Prolonged firing	1 round per howitzer per minute

b. **General Data Pertaining to the 155-mm Howitzer Carriage M1.**

Weights:	
Carriage, complete with weapon (without howitzer cover and accessories)	11,966 lb
Weapon	3,825 lb
Recoil Mechanism M6	1,582 lb
Spade	184 lb
Firing jack float	80 lb
Wheel, complete with tire and tube	409 lb
Right shield	79 lb
Left shield	86 lb
Required to lift off the ground:	
Coupled trails, for limbering to prime mover	552 to 371 lb
One trail without spade	349 to 198 lb
One trail with spade	390 to 249 lb
Dimensions:	
Width of track, center to center of wheels	82 in.
Maximum width, traveling position (outside walls of tires)	97 3/8 in.
Maximum height, traveling position (top of right shield)	71 in.
Height of center line of bore from ground (at zero elevation)	50 in.
Length of howitzer and carriage, limbered	26 1/2 ft
Road clearance (bottom of front trail stop)	14 in.
Maneuvers:	
Elevation	1,156 mils (65 deg)
Movement in elevation for one turn of handwheel	14.8 mils (50.2 min)
Traverse to right or left of mid-position	471 mils (26 1/2 deg)
Movement in azimuth for one turn of handwheel	9.6 mils (32.4 min)

TM 9-331

4

155-MM HOWITZER M1 AND 155-MM HOWITZER CARRIAGE M1

Figure 4 — 155-mm Howitzer M1 and 155-mm Howitzer Carriage M1 — Limbered to Prime Mover

TM 9-331
5-6

CHAPTER 2

HOWITZER AND CARRIAGE

Section I

DESCRIPTION AND FUNCTIONING OF HOWITZER

	Paragraph
General	5
Barrel assembly	6
Breech mechanism	7
Firing mechanism	8
Breech mechanism functioning	9
Firing mechanism functioning	10

5. GENERAL.

a. The 155-mm Howitzer M1 (fig. 5) consists of an alloy steel tube screwed into a breech ring and locked in place by a screw which is inserted through the forward upper wall of the breech ring.

b. The howitzer is supported and slides in recoil and counterrecoil on smoothly finished bearing surfaces, forward of the breech ring. Rotation of the tube in the mount is prevented by 2 keys on the upper right surface of the tube.

c. The breechblock is of the stepped-thread interrupted screw type, designed for use with separate loading ammunition. It is supported on a carrier, hinged to the right side of the breech ring. The breech is operated manually. The carrier is counterbalanced by a spring and cylinder type counterbalance.

d. The howitzer is fired by the percussion type Firing Mechanism M1, removably screwed into the firing mechanism housing at the rear of the breechblock. The firing mechanism is actuated by a lanyard-operated percussion hammer.

e. The howitzer designation (fig. 6), serial number, weight, name of manufacturer, and year of manufacture are stamped on the upper surface of the breech ring.

6. BARREL ASSEMBLY.

a. Tube.

(1) The tube (fig. 5) is formed in one piece. The exterior of the breech end is threaded to receive the breech ring. Immediately forward of the thread is a tapered recess to receive the breech ring locking screw.

(2) Except for the threaded portion which screws into the breech ring, the exterior of the tube is smoothly finished to form the recoil slide

TM 9-331

155-MM HOWITZER M1 AND 155-MM HOWITZER CARRIAGE M1

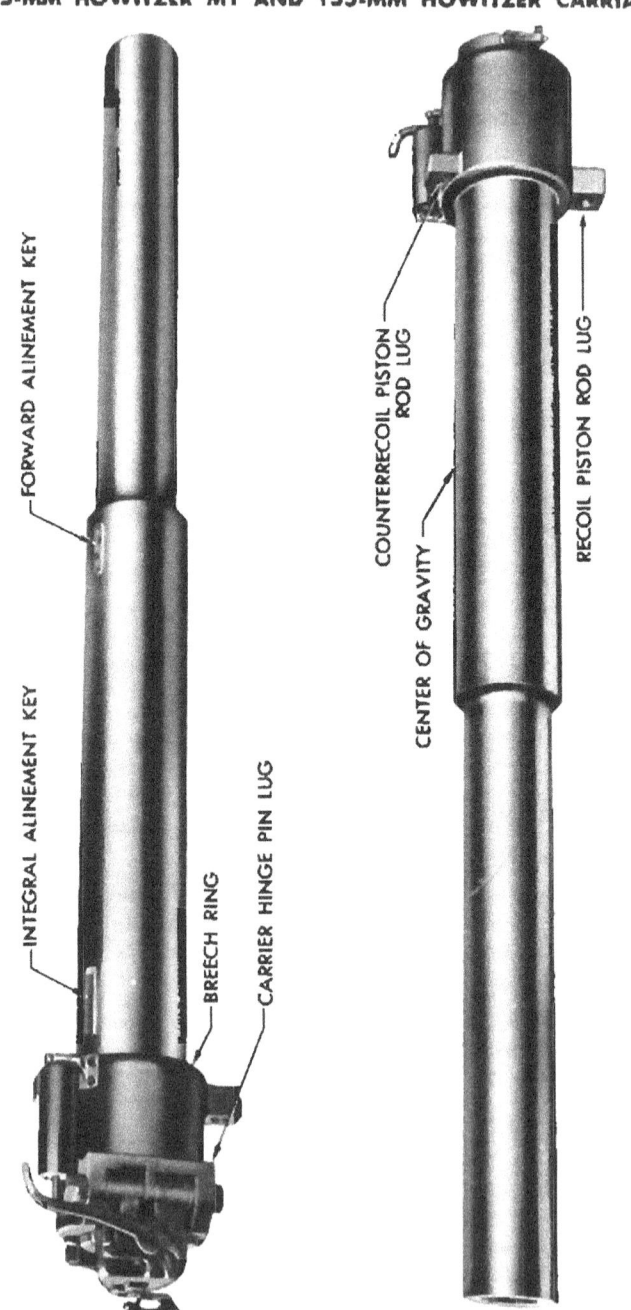

Figure 5 — 155-mm Howitzer M1 — Left (above) and Right (below)

TM 9-331
6

DESCRIPTION AND FUNCTIONING OF HOWITZER

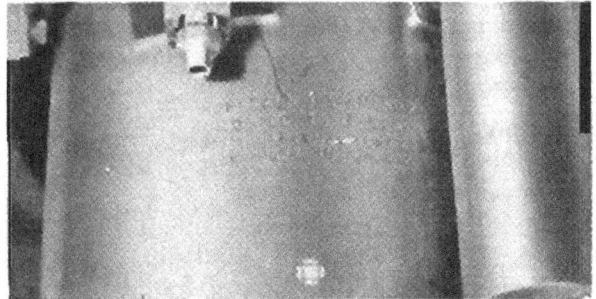

RA PD 54855

Figure 6 — Howitzer Designation and Weight Marking

which provides a sliding bearing in the mount. The forward section of this bearing surface is smaller in diameter than the rearward section, the change in diameter forming a shoulder between the bearings. The tube is not tapered.

(3) An integral longitudinal key (fig. 5) is formed in the upper right surface near the rear end of the rearward bearing. A similarly shaped key is screwed to the upper right surface of this bearing near its forward end.

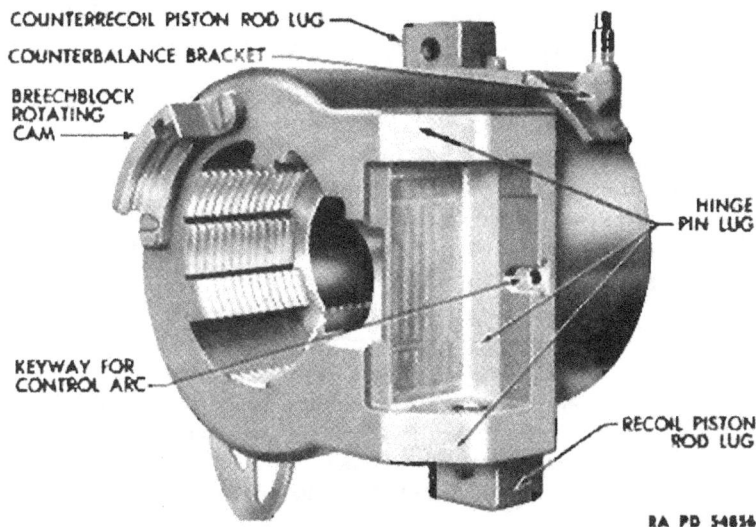

RA PD 54856

Figure 7 — Breech Ring of 155-mm Howitzer M1

155-MM HOWITZER M1 AND 155-MM HOWITZER CARRIAGE M1

The purpose of these keys is to maintain the alinement of the howitzer in the mount.

(4) The rear end of the bore is tapered to form the powder chamber. From the chamber to the muzzle, the bore is rifled with 48 grooves with a uniform right-hand twist of one turn in 25 calibers.

b. Breech Ring.

(1) The exterior of the breech ring (fig. 7) is cylindrical in general form. At the front end are lugs on the top and bottom for attachment of the howitzer to the counterrecoil and recoil piston rods. At the rear on the right side is the breechblock carrier hinge lug. Top and bottom ears on this lug are bored vertically to receive the carrier hinge pin. A small pin in the lower ear prevents rotation of the hinge pin driving washer bearing disk. The hinge pin lug is machined to receive and support the control arc which holds the breechblock alined, through the operating handle, to reenter the threaded sectors of the breech ring as the breech is being closed.

(2) The counterbalance bracket is attached by 4 screws to the upper right side of the breech ring near the forward end. The breechblock rotating cam is attached by means of 2 screws and a dowel pin to the upper left breech face of the breech ring. Two leveling plates for the gunner's quadrant are inlaid in the top surface of the breech ring.

(3) The forward portion of the breech ring bore is threaded to receive the threads on the tube. The rear portion of the bore is divided into 12 stepped sectors. Nine of these are threaded and three are plain to correspond with the exterior of the breechblock.

7. BREECH MECHANISM.

a. Breechblock.

(1) The breechblock (fig. 8) is of the cylindrical, stepped-thread, interrupted screw type. There are 9 threaded sectors and 3 plain sectors. The arrangement of the threads permits the breechblock to be locked or unlocked by its being rotated approximately one-tenth of a revolution, or 38 degrees.

(2) A large integral lug on the rear portion of the breechblock contains a groove, a cam slot, and a recess. The groove receives the crosshead by means of which the crankshaft rotates the breechblock. The crankshaft roller travels in the cam slot and moves the breechblock back and forth on the driver as the breechblock is being unlocked and locked. The safety latch seats in the recess when the Firing Mechanism M1 is screwed into place and locks the breechblock against rotation, and therefore against opening.

(3) The breechblock has a central stepped section from front to rear. The forward portion of this bore fits the obturator spindle. The central

DESCRIPTION AND FUNCTIONING OF HOWITZER

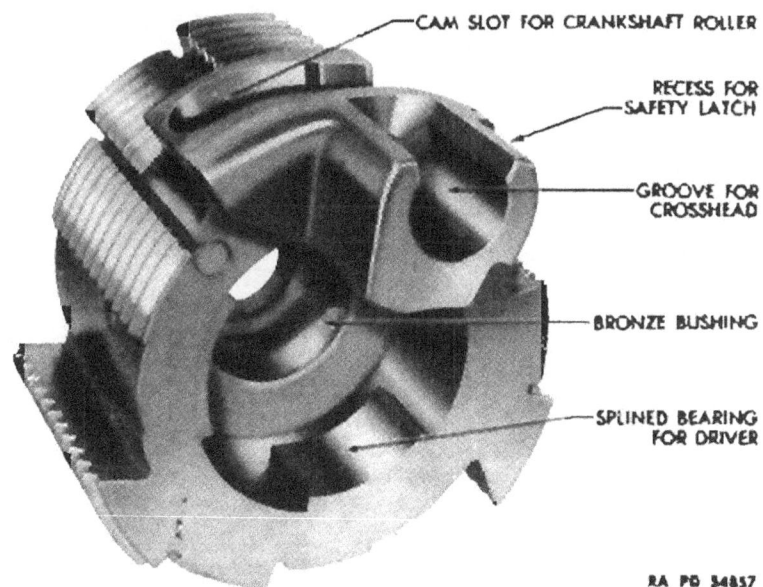

Figure 8 — Breechblock from Rear

portion of the bore is provided with a bronze bushing which forms the bearing surface on which the breechblock is carried on the hub of the carrier.

(4) The rear portion of the bore contains 3 wide splines which mate with 3 similar splines of the driver. They permit the breechblock to move backward and forward on the driver as the breechblock is being unlocked and locked, but do not permit the breechblock to turn independently of the driver.

b. **Obturator.**

(1) The obturating parts (fig. 9) consist of the obturator spindle with obturator spindle bushing, plug, and plug gasket, front and rear split rings, inner ring, gas check pad, filling-in disk, and obturator spindle spring.

(2) The rings, pad, and disk are assembled under the head of the obturator spindle on the front face of the breechblock and serve to seal the rear of the powder chamber against the rearward escape of powder gases when the howitzer is fired. The obturator spindle is drilled centrally from front to rear to permit flame from a primer, seated in the plug at the rear end of the spindle, to enter the powder chamber of the howitzer and ignite the charge to fire the howitzer.

155-MM HOWITZER M1 AND 155-MM HOWITZER CARRIAGE M1

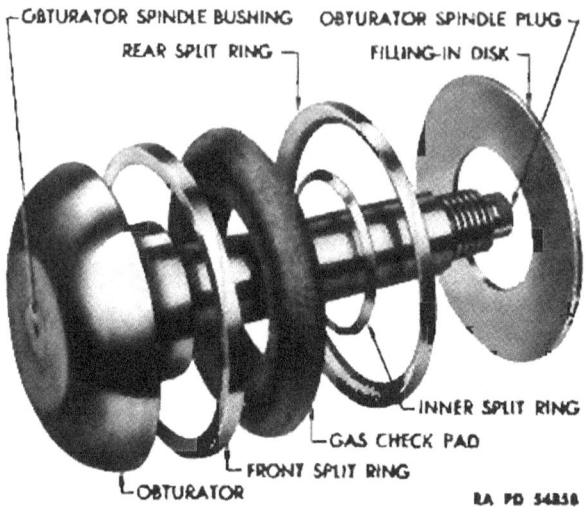

Figure 9 — Obturating Parts

(3) The obturator spindle (fig. 9) has a mushroom-shaped head and a cylindrical stem which fits into the bore of the obturator spindle sleeve. The rear surface of the head is slightly coned to fit the forward surfaces of the front split ring and the gas check pad, against which the head bears, and a shoulder which fits into a recess at the front of the breechblock.

(4) The rear end of the obturator spindle stem is threaded externally to receive the firing mechanism housing by which it is retained in the breechblock. A keyway, cut in the stem, is engaged by a key in the sleeve. The sleeve is keyed to the carrier and prevents rotation of the spindle. The obturator spindle plug, which is threaded into the rear end of the stem, is taper-bored to receive the primer case. The plug gasket protects the plug threads from powder gases.

(5) The gas check pad (fig. 9) is a slightly pliable pressed ring consisting principally of asbestos and binder. It is shaped and constructed to expand against the chamber wall and the stem of the obturator spindle when the head of the spindle is forced against the pad by powder pressure. The rings protect the edges of the pad at junctions with the chamber (gas check seat) and the obturator spindle. The filling-in disk is a flat, circular plate of the proper thickness to insure seating of the pad and rings against the tapered wall of the powder chamber.

(6) The obturator spindle spring functions inside the breechblock bore around the obturator spindle stem. The spring is seated against the

DESCRIPTION AND FUNCTIONING OF HOWITZER

rear end of the sleeve. It bears to the rear on the adapter to draw the obturator spindle, the gas check pad, and allied parts to a tight seat on the face of the breechblock.

(7) The obturator spindle sleeve has a smooth bore which supports the stem of the obturator spindle and a stepped exterior which fits the bore of the carrier. On the right side, the rear portion carries a key which fits a keyway in the carrier. A key extending into the bore at the rear fits a keyway in the stem of the obturator spindle. These keys and keyways prevent the obturator spindle and sleeve from rotating in the bore of the carrier.

c. **Breechblock Carrier.**

(1) The breechblock carrier (fig. 10) is a relatively thin-walled hinge member which supports the breechblock and its actuating parts. The right end of the carrier is supported between the upper and lower ears of the hinge pin lug and is bored vertically to receive the hinge pin about which it swings. A lug with a stop surface for the operating lever is located on the right side of the carrier.

(2) A cylindrical section with 2 outer bearing surfaces extends from the front of the left end of the carrier. The front surface of this cylindrical section provides a bearing for the breechblock while the rear surface forms a bearing for the driver. The forward end of the rear section is threaded to receive the driver retaining ring. The ring is locked with a lock screw. The bore through the cylindrical section receives the sleeve, obturator spindle, spring, adapter, and firing mechanism housing.

(3) The carrier houses the crankshaft and related breechblock operating parts. A transverse bore through the right wall is provided for the crankshaft bushing. The firing mechanism safety latch is mounted on the outside rear face of the carrier. The operating lever latch is mounted on the upper right surface of the carrier.

d. **Breechblock Actuating Mechanism.**

(1) The crankshaft (fig. 10) is a cylindrical shaft with 2 crank arms at one end. One arm carries a cylindrical bearing fitting the bore of the crosshead. The other arm has an integral pin which carries the crankshaft roller and washer. The crankshaft extends from the right side of the breechblock through the right wall of the carrier. It is supported in the carrier by the crankshaft bushing. Its right end is keyed to the interior of the bushing.

(2) The crosshead (fig. 10) has a cylindrical exterior bearing surface which slides in a corresponding groove in the breechblock. It has a bore at right angles to its bearing surface into which the crank arm projects to actuate the crosshead. The crosshead serves to rotate the breechblock in locking and unlocking.

TM 9-331

155-MM HOWITZER M1 AND 155-MM HOWITZER CARRIAGE M1

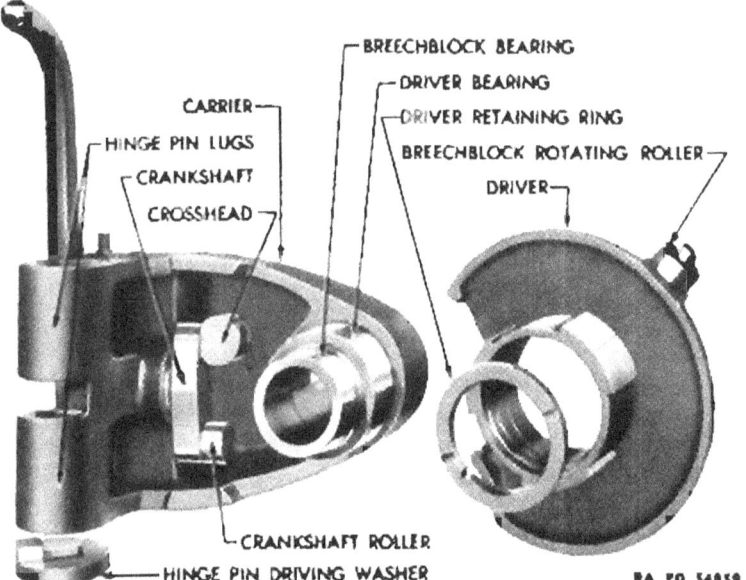

Figure 10 — Breechblock Carrier and Driver Viewed from Front

(3) The crankshaft roller (fig. 10) travels in a cam slot in the breechblock and moves the breechblock back and forth in its splined bearing on the driver.

(4) The crankshaft bushing (fig. 11) is a plain cylindrical bushing. It supports the crankshaft and fits the bore in the right side of the carrier. It is keyed to the crankshaft. The outer end of the bushing has an increased diameter, forming a shoulder which bears against the outer surface of the carrier. This extended hub is slotted at right angles to the crankshaft to receive the operating lever, which is attached to the end of the crankshaft by means of a screw.

(5) The driver (fig. 10) is a sleeve with a wide outside circular flange. A portion of the right side of the flange is removed to clear the rear lug of the breechblock. The driver is rotatably mounted on its bearing surface on the front of the carrier and is retained in position by the driver retaining ring. The rear of the driver sleeve lies against a shoulder on the carrier.

(6) The forward portion of the driver sleeve extends into the recess in the rear of the breechblock and is splined to the breechblock. The breechblock can be moved forward and backward independently of the driver, but both are made to rotate as a unit.

TM 9-331
7

DESCRIPTION AND FUNCTIONING OF HOWITZER

(7) The block rotating roller (fig. 10) is mounted on a pin set in a lug on the upper left edge of the driver flange.

(8) The block rotating cam (fig. 12) is an arc-shaped bracket attached to the upper left rear face of the breech ring by means of a dowel and 2 screws. The block rotating cam is grooved in a curved path to guide the block rotating roller when the breechblock is near the closed position.

(9) The operating lever (fig. 11) is a flat-sided lever with a bent tubular hand grip. Its lower end is inserted through the slot in the crankshaft bushing to which it is secured by means of a screw which extends into the right end of the crankshaft.

(10) The control arc (fig. 11) is a smooth-topped sector mounted midway between the upper and lower ears of the hinge pin lug. It is held in place by a countersunk screw. An integral projecting finger on the lower end of the operating lever rides on the upper surface of the control arc to prevent rotation of the lever and crankshaft when the carrier is in open position.

e. **Hinge Pin.**

(1) The hinge pin (fig. 11) hinges the carrier to the breech ring. It consists of a cylindrical body with an arm extending outwardly from its head (fig. 12). An integral pin extending from the top of the arm is provided for the attachment of the counterbalance tension rod.

(2) The hinge pin extends downward through the hinge pin lug and carrier and is retained by the hinge pin collar and detent. The hinge pin is squared near its lower end to fit the square hole in the hinge pin driving washer (figs. 10 and 11). This washer is keyed to the bottom of the carrier and thus forces the hinge pin and carrier to swing as a unit.

(3) The flat driving washer bearing disk (fig. 11), under the driving washer, provides supporting bearing surface for the driving washer and carrier. The bearing disk is prevented from rotating by a pin in the lower ear of the hinge pin lug.

f. **Counterbalance.**

(1) The counterbalance (fig. 12) is a cylinder with its ends closed by cylinder heads and containing a heavy compression spring. This spring is compressed between the rear cylinder head and the counterbalance piston, a disk which fits the cylinder freely. It is compressed by the counterbalance tension rod, which connects the counterbalance piston with the arm of the carrier hinge pin.

(2) The counterbalance facilitates closing the breech when the howitzer is elevated, under which condition gravity opposes the swinging of the breechblock into the breech recess. It also tends to hold the carrier

TM 9-331
7
155-MM HOWITZER M1 AND 155-MM HOWITZER CARRIAGE M1

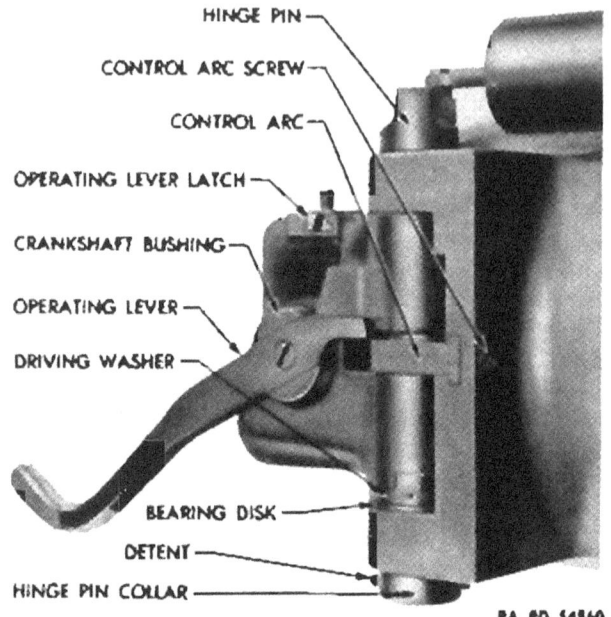

Figure 11 — Carrier Hinge Pin and Control Arc

in open position by means of spring tension when the counterbalance tension rod and hinge pin arm pass dead-center position.

(3) The mechanism is supported and hinged on the counterbalance bracket which is mounted on the forward right side of the breech ring. It is attached to the arm of the hinge pin by means of the counterbalance tension rod. The rod end has an eye of keyhole shape to permit assembly and removal over the head of the pin.

g. **Operating Lever Latch.**

(1) The operating lever latch (fig. 12) latches the lever in fully raised position, thereby locking the breech in closed position. It is a small plunger which fits slidably into a shouldered groove in the upper right edge of the carrier.

(2) The latch is spring-loaded to keep it extended except when it is being manually retracted. A round knurled handle is screwed through the latch and into a recess in the carrier. The recess limits the movement of the latch to the right and left, holding the latch in the carrier.

h. **Safety Latch.**

(1) The safety latch (fig. 12) prevents the breech from being opened before the firing mechanism has been unseated. It also prevents the

TM 9-331

DESCRIPTION AND FUNCTIONING OF HOWITZER

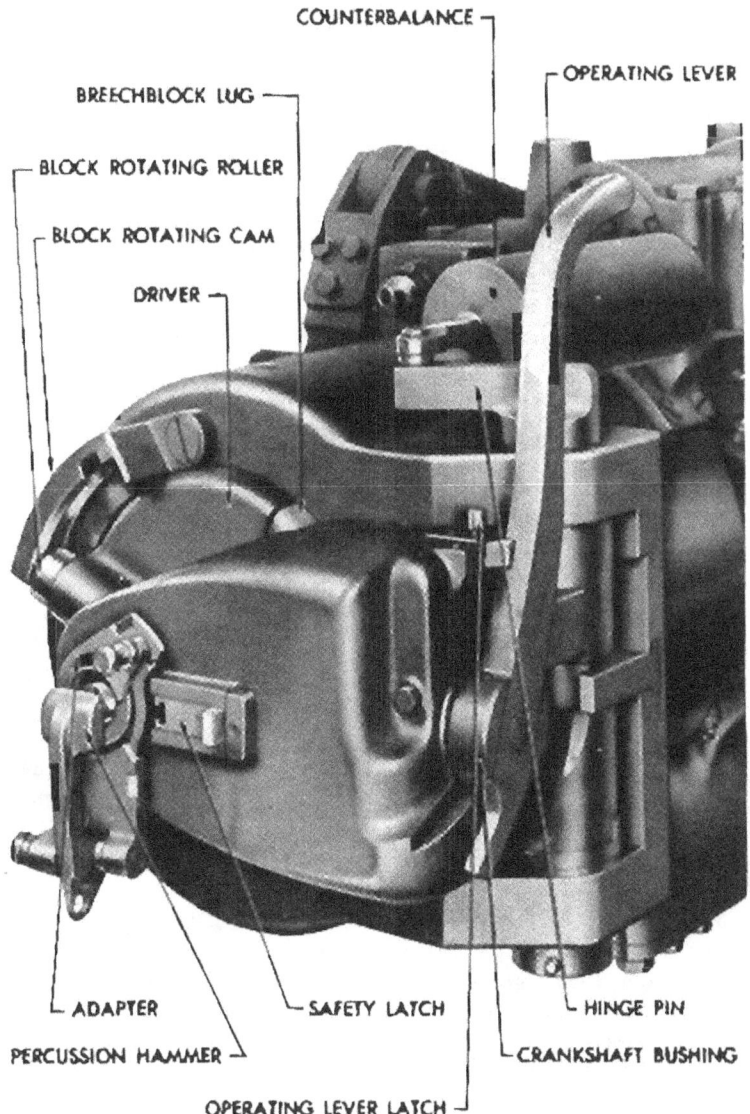

Figure 12—Breech Mechanism of the 155-mm Howitzer M1

155-MM HOWITZER M1 AND 155-MM HOWITZER CARRIAGE M1

firing mechanism from being seated before the breechblock has been returned to closed and locked position.

CAUTION: While it is possible to open or close the breech with the firing mechanism partially removed or inserted, this practice is strictly prohibited.

(2) The safety latch is an L-shaped piece with a knurled, rectangular handle. One leg is tongued to slide from side to side in grooves in lugs on the rear of the carrier. The other leg extends forward in the carrier bore and is grooved on the left side for the safety latch plunger which slides forward and backward in this groove. The front end of this leg of the latch enters a recess in the breechblock when the latch is forced to the right and the breechblock is in locked position. The plunger has an angular point which projects to the left into openings in the adapter and firing mechanism housing.

(3) The latch and plunger are drilled to receive a small pin which is mounted in the carrier and extends to the left. The pin and holes do not normally mate. When the mechanism is to be disassembled, the latch and plunger must be maneuvered until the pin enters the holes in the latch and plunger. This permits the latch to be moved sufficiently to the right to clear the plunger from its holes in the adapter and the firing mechanism housing.

(4) The right end of the latch is drilled to receive the latch spring, which presses the latch to the left. Movement of the latch to the right is confined by the safety latch plate which is screwed to the rear face of the carrier at the right ends of the safety latch lugs.

(5) With the breechblock locked and the Firing Mechanism M1, seated in its housing, the safety latch plunger is held to the right by the Firing Mechanism M1, pressing it back into the opening in the housing. This forces the safety latch to the right and engages the front end of the latch with the recess in the breechblock, locking the breechblock against rotation.

(6) When the Firing Mechanism M1 is removed, the latch and plunger move to the left under spring pressure. The latch frees the breechblock, and the plunger projects into the firing mechanism housing sufficiently to prevent replacement of the Firing Mechanism M1. Rotation of the breechblock to unlock moves the recess out of alinement with the latch. This locks the latch in the leftward position until the breechblock is rotated back to locked position.

8. FIRING MECHANISM.

a. General. The firing mechanism consists not only of the Firing Mechanism M1, which is removed and replaced as a unit between the firing of successive rounds, but also of the adapter and the firing mecha-

DESCRIPTION AND FUNCTIONING OF HOWITZER

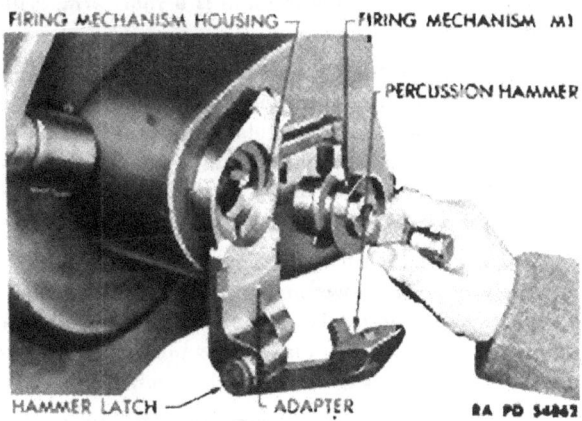

Figure 13—Firing and Percussion Mechanism

nism housing which retain the firing mechanism in the breech, and the percussion mechanism which actuates the Firing Mechanism M1.

b. Adapter.

(1) The adapter (fig. 13) is a sleeve, shouldered externally and internally, which fits into the rear end of the breechblock bore of the carrier and into the rear of the breechblock. It has a flat, cam-shaped upper arm, recessed to retain the plunger of the Firing Mechanism M1, and a rectangular lower arm which supports the percussion mechanism.

(2) An integral key on the left side of the adapter engages a slot in the bore of the carrier and prevents the adapter from rotating in the carrier. The adapter is retained in the carrier by the firing mechanism housing which fits the interior of the adapter and screws onto the rear end of the stem of the obturator spindle.

(3) A slotted hole in the right side of the adapter mates with a similar hole in the right side of the firing mechanism housing and permits passage of the safety latch plunger into the housing to enable the safety latch to function.

c. Percussion Mechanism.

(1) The percussion mechanism (fig. 13) consists of the hammer and hinge pin, and the hammer latch. These are mounted on the lower arm of the adapter. When a lanyard is attached to the hammer and pulled, the hammer swings in an arc and strikes the firing pin of the Firing Mechanism M1, firing the piece.

(2) The percussion hammer (fig. 13) is a lever with a bored hub which fits into the fork of the lower arm of the adapter where it is hinged

on the hammer hinge pin. Below the hub is a short arm, drilled for the attachment of a lanyard.

(3) The upper arm of the hammer terminates in a rectangular head with a raised, rounded striking surface formed on its flat front face. Below the striking surface is a rectangular projection positioned to strike the rim of the firing mechanism block if the Firing Mechanism M1 has not been screwed home. This is a safety feature to prevent impact on the firing pin before the primer case has been properly chambered.

(4) The hammer latch pin (fig. 13) and spring are housed in a bore in a rearward extension of the lower arm of the adapter. Movement of the pin is controlled by the hammer latch knob; the pin is pressed toward the hammer by means of the spring. The knob is provided with a small positioning pin which may be engaged to lock the latch in different positions.

(5) When the hammer latch is released, the latch pin protrudes in the path of the hammer and prevents its being raised. When the knob is drawn to the left and turned, the latch pin is held from the path of the hammer and the hammer is permitted to swing in its arc. When the hammer is raised, the knob can be turned until the latch pin enters a recess in the hammer, holding the hammer in an inoperative, upright position.

d. Firing Mechanism Housing. The firing mechanism housing (fig. 13) is a shouldered cylinder which fits the interior of the adapter. It has a stepped central bore, threaded in front to screw onto the obturator spindle and threaded in the rear to receive the Firing Mechanism M1. A slotted hole in the right-hand wall provides for passage of the safety latch plunger into the threaded space for the Firing Mechanism M1.

e. Firing Mechanism M1.

(1) The Firing Mechanism M1 (figs. 14 and 15) consists of the firing pin, primer holder, and related parts, housed in the firing mechanism block. The mechanism screws, as a unit, into the firing mechanism housing in the rear end of the breechblock. It is removed and replaced between the firing of successive rounds.

(2) The firing mechanism block (fig. 16) is a short, flanged cylinder with a rearward rim extending around a major portion of the flange. This rim prevents the percussion hammer from striking the firing pin unless the mechanism is screwed into firing position. An arm integral with the flange extends to the side at the rear of the block. At the end of this arm is a spring-loaded plunger and handle by means of which the mechanism is latched in firing position and manipulated in removal and replacement. The block has a coarse thread on its outside, and its bore is threaded in front to receive the primer holder and in the rear to receive the firing pin housing.

DESCRIPTION AND FUNCTIONING OF HOWITZER

Figure 14—Firing Mechanism M1—Front

Figure 15—Firing Mechanism M1—Rear

(3) The firing pin (fig. 16) has a shouldered, cylindrical body with a rounded rear contact surface and a flat-ended, pinlike nose. The firing pin is held in the bore of the block by the firing pin housing at the rear and by the firing pin spring, guide, and primer holder at the front. The firing pin housing is a bushing which screws into the rear of the block. Its bore form a bearing for the rear portion of the firing pin. The firing pin guide (fig. 16) is a cup which fits, closed and forward, in the forward end of the bore of the block. A hole in the closed end supports the point of the firing pin. The firing pin spring is compressed between the closed end of the guide and the shoulder of the pin.

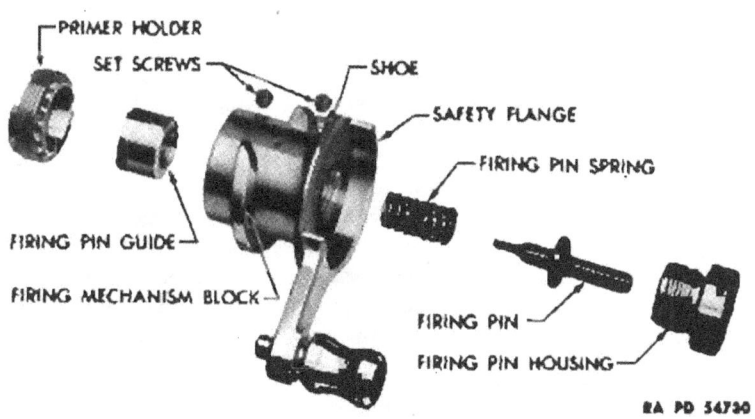

Figure 16—Exploded View of Firing Mechanism M1

TM 9-331
8-9

155-MM HOWITZER M1 AND 155-MM HOWITZER CARRIAGE M1

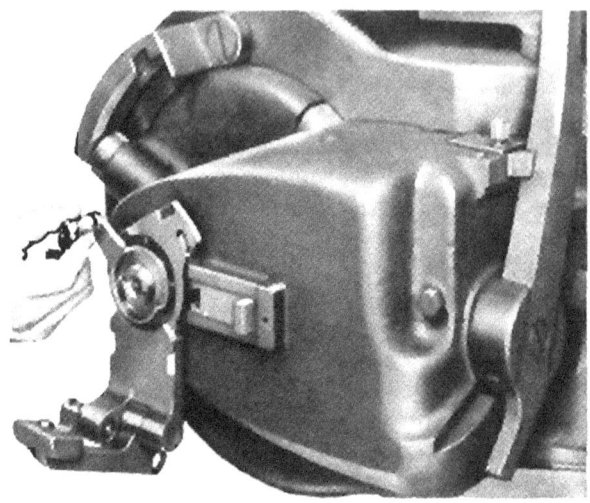

Figure 17 — Removing Firing Mechanism M1

(4) The primer holder (fig. 16) is ring-shaped. It screws into the front of the block, holding the guide and spring in place. Its flanged front has a U-shaped slot to receive the head of the primer case with the cap of the case directly in front of the nose of the firing pin. The primer holder is locked in the block by a set screw which engages one of a number of notches machined in the rear outer edge of the holder. The firing pin housing also is locked by a set screw and a copper shoe (fig. 16) is inserted ahead of the screw to prevent damage to the threads of the housing. Holes are provided in the exposed heads of the holder and housing for use of a spanner wrench in assembly and disassembly.

9. BREECH MECHANISM FUNCTIONING.

a. **General.** The functioning of the breech mechanism, either in opening the breech or in closing it, comprises the following two cycles of operation: unscrewing the interrupted threads of the breechblock and breech recess; swinging the breechblock out of the breech recess. Operation of the breech is by hand, opening and closing being accomplished by means of the operating lever at the right side of the breech.

b. **Opening the Breech.**

(1) FIRST CYCLE. When the breech is closed and the firing mechanism is seated in firing position, the front end of the safety latch is seated in a recess in the rear of the breechblock, preventing rotation of the breechblock. The breechblock is released for rotation by removing

TM 9-331
9

DESCRIPTION AND FUNCTIONING OF HOWITZER

Figure 18 — Releasing Operating Lever Latch

the firing mechanism (fig. 17). This permits the safety latch to slide to the left, and out of engagement, under pressure of its spring.

(2) The operating lever latch must be moved to the left (fig. 18) to release the operating lever from its upright position. When the lever is drawn rearward and downward, it rotates the crankshaft bushing and the crankshaft inside the carrier. This causes the crosshead on one of the arms of the crankshaft to force the breechblock to rotate clockwise and disengage the interrupted threads. Disengagement of the threads requires but one tenth of a revolution of the breechblock.

Figure 19 — Unsealing Breech

TM 9-331
9

155-MM HOWITZER M1 AND 155-MM HOWITZER CARRIAGE M1

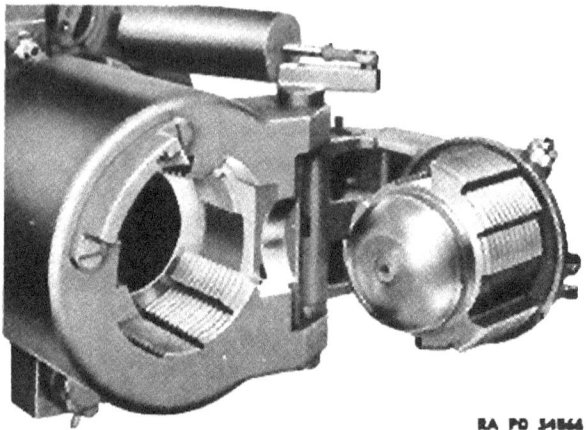

RA PD 34864

Figure 20 — Breech Mechanism Open

(3) As the threads disengage and the crankshaft continues to turn, the roller on the opposite arm of the crankshaft follows the cam groove in the breechblock lug to move the breechblock rearward in the carrier (fig. 19). This rearward movement starts withdrawal of the obturator spindle head from the powder chamber and places the breechblock and spindle in position to clear the breech ring as they are swung rearward with the carrier.

(4) Rotation of the breechblock is stopped in proper alinement by the contact of the operating lever with its stop lug on the carrier.

(5) SECOND CYCLE. As the operating lever is swung to the right and forward, the breechblock and carrier are swung about the axis of the carrier hinge pin until stopped in the open position against the breech ring (fig. 20). Clearance cuts are provided on the breech ring and breechblock to permit the block to swing freely.

(6) As the carrier swings away from the breech, the projecting finger on the hub end of the operating lever slides on the top surface of the control arc. This maintains the breechblock in the proper position to enter the breech ring when the carrier is swung back to closed position.

(7) The change from rotating movement of the breechblock to swinging motion of the carrier is controlled by the block rotating roller which is mounted on the flange of the driver. The driver rotates as a unit with the breechblock on the carrier. The roller travels in the curved cam groove of the block rotating cam which is mounted on the left upper rear face of the breech ring. The roller leaves the cam groove after the breechblock is locked in position by the control arc.

DESCRIPTION AND FUNCTIONING OF HOWITZER

(8) The counterbalance assists in opening and closing the breech when the howitzer is elevated, and tends to hold the carrier in open position. As the breech is swung open, the driving washer of the hinge pin forces the hinge pin to turn with the carrier. The counterbalance tension rod, which is connected to the arm on top of the hinge pin, is drawn outwardly until the hinge pin arm passes dead-center position. The compression of the spring in the counterbalance resists the opening movement of the carrier and stores energy to assist in closing. When the arm passes dead-center position, spring tension tends to hold the carrier in open position.

c. Closing the Breech.

(1) FIRST CYCLE. As the operating lever is drawn backward and to the left, the carrier is swung to closed position. When the howitzer is elevated, this movement is assisted by the energy stored up in the compressed spring of the counterbalance. The projecting finger on the hub end of the operating lever applied against the top surface of the control arc maintains the breechblock in the proper position to enter the breech recess, and the block rotating roller enters the groove of the block rotating cam.

(2) SECOND CYCLE. As the lever is raised upward and forward, the crankshaft is rotated and the crankshaft roller and crosshead draw the breechblock forward in the breech recess and rotate the block counterclockwise to engage the interrupted threads. As the operating lever is pressed fully forward to lock the breech, the lever is latched by the operating lever latch.

(3) When the Firing Mechanism M1 is screwed into its housing, the safety latch is pressed to the right against the tension of its spring. This forces the front end of the safety latch to seat in the breechblock and prevents the breechblock from being rotated.

10. FIRING MECHANISM FUNCTIONING.

a. In order for the firing mechanism to function, the breach must be fully closed, a percussion type primer must be inserted in the primer holder in the front end of the firing mechanism, and the firing mechanism must be screwed into the firing mechanism housing. Firing is accomplished by a sharp pull on the lanyard, which swings the percussion hammer to strike the firing pin.

b. When the rear contact surface of the firing pin is struck by the percussion hammer, the pin is driven forward against the tension of the firing spring. This causes the nose of the firing pin to protrude through the firing pin guide and to strike the cap of the primer seated in the primer holder, firing the piece.

155-MM HOWITZER M1 AND 155-MM HOWITZER CARRIAGE M1

c. If the Firing Mechanism M1 has not been fully screwed into the firing mechanism housing, the percussion hammer is prevented from striking the firing pin by the rim on the rear of the firing mechanism block which contacts the forward projecting lug on the hammer. When the Firing Mechanism M1 is screwed home, the projecting lug on the hammer passes through a gap in the rim of the block, allowing the hammer to strike the firing pin.

TM 9-331
11

Section II

DESCRIPTION AND FUNCTIONING OF CARRIAGE

	Paragraph
General	11
Recoil mechanism and cradle	12
Recoil mechanism functioning	13
Top carriage	14
Elevating mechanism	15
Traversing mechanism	16
Equilibrators	17
Bottom carriage	18
Firing jack	19
Traveling lock	20
Trails	21
Axle and hubs	22
Wheels and tires	23
Electric brakes	24
Hand brakes	25

11. GENERAL.

a. The 155-mm Howitzer Carriage M1 (fig. 21) is of the 2-wheel, split-trail type. The wheels are equipped with pneumatic tires. In traveling position, the trails are locked together by a toggle type clamping mechanism and are limbered directly to the prime mover. The carriage is equipped with electric power brakes.

b. The howitzer is fired from 3-point suspension, with the trails spread and the carriage resting upon an integral firing jack, the wheels being clear of the ground. The firing stresses are transmitted through the trunnion pins, top carriage, bottom carriage and trails to the spades which are attached to the rear ends of the trails. The spades, being buried in the ground, transmit the reaction back to the trails and prevent movement of the carriage.

c. Recoil energy is absorbed and the howitzer is returned to battery after firing by the Recoil Mechanism M6 of the hydropneumatic, variable recoil type. This recoil mechanism is of built-up construction and incorporates the cradle which supports the howitzer and in which the howitzer slides in recoil and counterrecoil. The tipping parts of the howitzer and carriage are supported on trunnion pins in the top carriage.

d. For ease of traversing, the top carriage is traversed on the pintle. Spring type equilibrators assist in elevating and depressing the muzzle-heavy weapon. A traveling lock is provided to stabilize the howitzer in

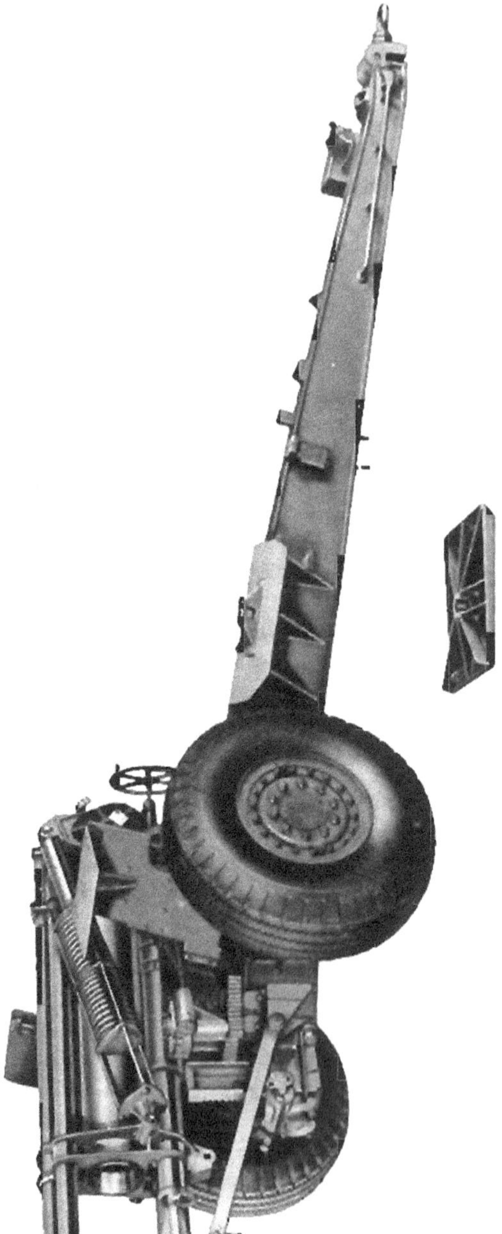

TM 9-331
11

155-MM HOWITZER M1 AND 155-MM HOWITZER CARRIAGE M1

Figure 21 — 155-mm Howitzer Carriage M1, Without Weapon — Left Front View

RA PD 54867

30

TM 9-331
11-12

DESCRIPTION AND FUNCTIONING OF CARRIAGE

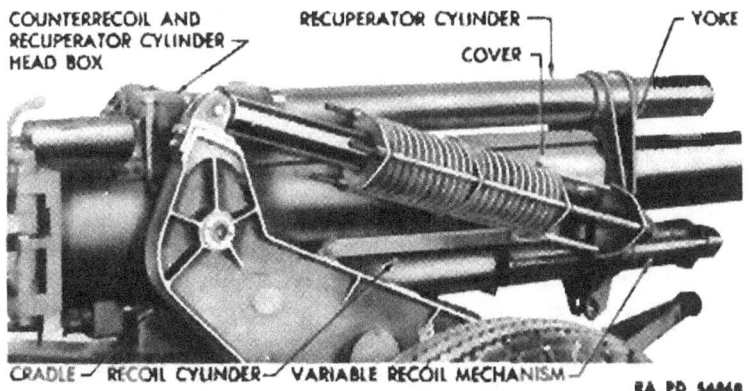

Figure 22 — Recoil Mechanism M6 — Right Side

traveling position. Shields are provided for the protection of the personnel. The brakes are provided with hand levers for parking and with a safety switch for automatic brake application in the event of a break-in-two from the prime mover. The wheels support the carriage on antifriction bearings.

12. RECOIL MECHANISM AND CRADLE.

a. **General.** The Recoil Mechanism M6 (fig. 22), of the 155-mm Howitzer Carriage M1, is of the hydropneumatic, variable recoil type. It is composed principally of the cradle, yoke, and cover, the recoil cylinder, variable recoil mechanism, replenisher, counterrecoil and recuperator cylinder head box, recuperator cylinder, counterrecoil cylinder, and their assembled parts.

b. **Purpose.**

(1) The purpose of the recoil portion of the recoil mechanism is to absorb and control the backward thrust of the weapon created by firing, and to check its movement in a manner so gradual as not to cause displacement of the carriage.

(2) The purpose of the counterrecoil and recuperator portions of the recoil mechanism is to store part of the recoil energy and, upon completion of the recoiling action, to return the howitzer into battery in order that it may be fired again. Buffer action controls the concluding movement of the howitzer as it returns to battery, permitting it to do so without appreciable shock to the mechanism.

TM 9-331

155-MM HOWITZER M1 AND 155-MM HOWITZER CARRIAGE M1

c. Cradle, Yoke, and Cover.

(1) The cradle (fig. 22) is a casting containing two bores, one above the other. It is supported on two-row roller bearings housed in the cradle trunnion bearings. The roller bearings rest on pins which are fitted in the top carriage trunnion bearings. The cradle trunnion bearings are reinforced by ribs.

(2) The larger bore of the cradle is equipped with a bronze liner which supports the howitzer and forms the bearing surface in which the howitzer slides in recoil and counterrecoil. The smaller bore, beneath the larger one, supports the rear end of the recoil cylinder.

(3) The counterrecoil and recuperator cylinder head box (fig. 22) is mounted on top of the cradle, and the elevating arc is mounted below the cradle. The front end of the arc is attached to the recoil cylinder by means of a bracket. The arc is held in alinement with the cradle by a key.

(4) The yoke (fig. 22) is a casting, containing bores which are alined with those in the cradle and the head box. It is located near the front ends of the recoil cylinders. It supports and holds in alinement the front ends of the howitzer and cylinders. It also supports the variable recoil mechanism and the replenisher.

(5) Pins in the sides of the yoke form trunnion attachments for the front ends of the equilibrators. A fork at the bottom of the yoke is drilled for a pin which holds the traveling lock in locked position.

(6) The cover (fig. 23) is of welded steel construction and connects the cradle and yoke on which it is held in position by shouldered bearings. It protects the portion of the bearing surface on the howitzer between the cradle and yoke. A wiper, held in place by a retaining washer (fig. 26) is mounted on the front of the yoke around the barrel and prevents dirt, water, and other foreign matter from getting inside the cover.

(7) At the upper right side, the interiors of the cradle liner and cover are fitted with guides which contain keyways in which the keys on the howitzer travel during recoil and counterrecoil to keep the howitzer from rotating in its mount.

d. Recoil Cylinder.

(1) The recoil cylinder (fig. 23) is located directly beneath the howitzer. It is threaded on its rear end to fit the mating bore in the cradle. The front end is supported by the yoke and is locked on the yoke by means of nuts.

(2) The recoil stuffing box assembly seals the rear end of the recoil cylinder and forms a guide for the recoil piston rod. The front end is sealed by the recoil cylinder head assembly which also forms the end bearing for the variable recoil control rod. The recoil cylinder houses

DESCRIPTION AND FUNCTIONING OF CARRIAGE

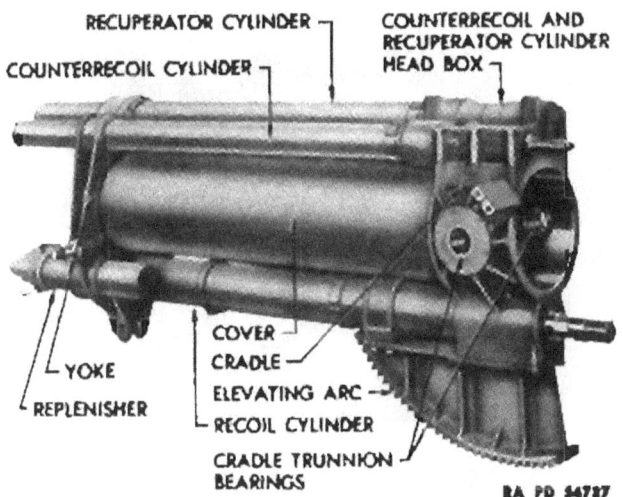

Figure 23 — Recoil Mechanism — Left Side — Removed from Carriage

the recoil piston and piston rod and the variable recoil control rod. All space in the recoil cylinder not occupied by the mechanism is filled with recoil oil.

(3) The rear end of the recoil piston rod is connected to the lower lug on the breech ring and travels backward with the weapon during recoil. The piston and rod are drilled centrally to provide a bearing in which the control rod, which does not travel backward with the weapon during recoil, slides during the recoil movement. The bore of the piston rod is of smaller diameter at its rear end to form the buffer chamber.

(4) The variable recoil control rod is mounted rotatably in the bore of the recoil cylinder head at the front end of the cylinder. A number of long grooves of varying depth are cut lengthwise in the surface of the control rod. In recoil, oil in the rear of the piston must pass through ports in the piston rod to these grooves, and out in front of the piston. This has a throttling action on the oil which absorbs a great portion of the recoil energy.

(5) The control rod is turned by a segment (fig. 24) at its front end which meshes with a segment actuated by the variable recoil mechanism. Turning of the control rod regulates the volume of oil which can pass through the grooves. This, in turn, varies the length of recoil of the weapon.

(6) The rear end of the control rod is of smaller diameter and is provided with throttling grooves to form a spear-head type buffer. Near the

155-MM HOWITZER M1 AND 155-MM HOWITZER CARRIAGE M1

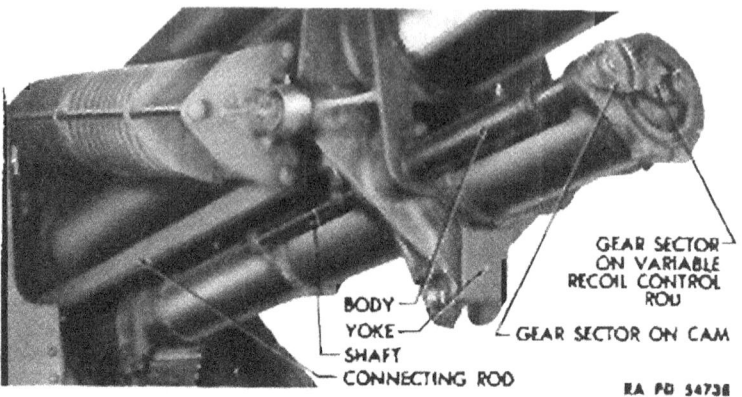

Figure 24 — Variable Recoil Mechanism

end of the counterrecoil movement, this buffer enters the buffer chamber of the recoil piston rod. The oil trapped in the buffer chamber must escape through the grooves, and, in doing so, cushions the final movement of the weapon as it returns to battery.

e. **Variable Recoil Mechanism.**

(1) The variable recoil mechanism (figs. 22 and 24) controls the length of recoil of the weapon at various elevations. The body of this mechanism is mounted in and on the right side of the yoke near the front end of the recoil cylinder. It houses a tubular cam which is machined with a spiral slot. A gear sector, mounted on the front end of the cam, engages a similar sector on the front end of the variable recoil control rod in the recoil cylinder.

(2) A shaft slides in the bore of the cam. This shaft is provided with a pin which engages the spiral slot in the cam (fig. 28). The shaft is connected to the top carriage by a connecting rod. As the weapon is elevated or depressed, the connecting rod moves the shaft forward or backward. This causes the pin on the shaft to turn the cam and its gear sector, which turns the gear sector on the control rod. The turning of the control rod regulates the opening of the oil passages in the recoil cylinder and determines the length of recoil.

f. **Replenisher.**

(1) The replenisher (figs. 23 and 25) serves as an oil reservoir for the recoil cylinder. It stores excess oil forced from the recoil cylinder by expansion of the oil due to increased atmospheric temperature or to heat developed by firing. It supplies oil required to compensate for contraction of the oil due to low temperatures.

TM 9-331
12

DESCRIPTION AND FUNCTIONING OF CARRIAGE

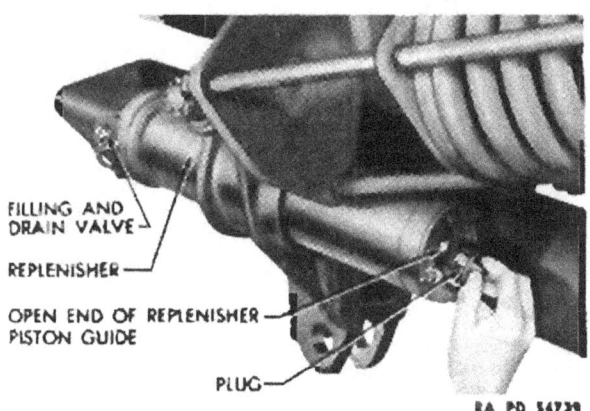

Figure 25 — Replenisher Showing Plug Removed

(2) The replenisher cylinder is mounted in a bore in the yoke at the left and near the front end of the recoil cylinder. A tube in the head of the replenisher connects the replenisher with the recoil cylinder. The replenisher is fitted with a spring-loaded piston. The piston, which has an oiltight packing arrangement, is forced rearward against spring pressure as oil enters the replenisher. The spring-loaded piston forces the oil into the recoil cylinder when oil is required.

(3) An extension on the rear of the piston slides in a guide which closes the rear end of the replenisher. The guide is open at the rear and provides a means for determining the amount of reserve oil in the replenisher. This reserve can be determined by inserting a scale in the open rear end of the guide and measuring the distance from the rear face of the guide to the rear end of the piston extension (par. 29 h). A plug (fig. 25) is provided to protect the open end of the guide from dirt and other foreign matter.

g. *Counterrecoil and Recuperator Cylinder Head Box.* This casting (fig. 23), which contains 2 bores, is bolted to the top of the cradle. The rear end of the counterrecoil cylinder is assembled into the front of the smaller bore at the left. The rear end of the recuperator cylinder is assembled into the front of the other bore. The inside diameters of

TM 9-331
155-MM HOWITZER M1 AND 155-MM HOWITZER CARRIAGE M1

Figure 26 — Front Ends of Recuperator and Counterrecoil Cylinders

Figure 27 — Rear Ends of Recuperator and Counterrecoil Cylinders

these 2 cylinders are continuous with the bores in the head box. The 2 bores are connected by an opening and house units of the counterrecoil and recuperator mechanisms.

h. **Counterrecoil Cylinder.**

(1) The counterrecoil cylinder (fig. 23) is the smaller of the 2 cylinders mounted above the howitzer. Its front end is closed by the respirator and is supported by, and is a slip fit in, a bore in the yoke. Its rear end is assembled in its bore in the counterrecoil and recuperator cylinder head box. The rear end of this bore is sealed by the counterrecoil stuffing box which forms the bearing for the counterrecoil piston rod.

(2) The counterrecoil piston and piston rod slide in the counterrecoil cylinder (fig. 28). This rod is connected to the top lug of the breech ring and travels backward with the weapon during recoil. The piston is fitted with an oiltight packing arrangement which permits it, when drawn backward in recoil, to force the oil in the cylinder to the rear and through the opening into the recuperator cylinder.

(3) The respirator (fig. 26) in the front end of the counterrecoil cylinder is equipped with a spring-loaded ball check valve. Its purpose is to release any air pressure in front of the counterrecoil piston when the piston is moved forward during counterrecoil.

i. **Recuperator Cylinder.**

(1) The recuperator cylinder (fig. 23) is the larger of the 2 cylinders mounted above the weapon. Its front end is supported in a bore in the yoke and is locked by 2 nuts. Its rear end is assembled in the

DESCRIPTION AND FUNCTIONING OF CARRIAGE

front of the right bore of the counterrecoil and recuperator cylinder head box. The front end is sealed by the recuperator front cylinder head which is equipped with a gas (nitrogen) charging valve and protected by a cover (fig. 26). The rear end of the recuperator cylinder bore of the head box is sealed by the recuperator rear cylinder head. This head is fitted with an oil filling valve and plug and the oil index (fig. 27).

(2) The recuperator cylinder houses the floating piston (fig. 28). The floating piston is a grease seal which separates the oil at the rear end of this cylinder from the compressed nitrogen at the front end. This piston moves forward or backward, depending upon the direction from which the greater pressure comes. The seal is obtained by grease held tightly packed by spring tension between 2 disks. Both disks are fitted with elastic packing held tightly against the cylinder walls by the tension of Belleville springs.

(3) The regulator valve is housed in the recuperator cylinder bore of the head box. It permits free passage of oil from the counterrecoil cylinder to the recuperator cylinder during recoil of the howitzer but regulates the flow of oil back into the counterrecoil cylinder during counterrecoil.

(4) The oil index (fig. 27) indicates the presence of an oil reserve in the recuperator (par. 29 c), which is necessary for proper functioning of the counterrecoil mechanism. When no reserve is present, the rear end of the floating piston presses on the oil index rod (fig. 29), which is racked through a gear to the oil index. This draws the rear end of the oil index below the outside face of the cylinder head. The weapon should not be fired when the index is in this position.

13. RECOIL MECHANISM FUNCTIONING.

a. General. The recoil mechanism, located below the howitzer, and the counterrecoil mechanism, located above the howitzer, operate in conjunction with one another to control both the recoil and the counterrecoil of the weapon. The only direct connection between the 2 mechanisms is the breech ring to which both are individually attached.

b. Recoil Action.

(1) When the howitzer is fired, the howitzer recoils in the cradle, and the recoil and counterrecoil pistons are drawn backward through their cylinders by their piston rods, which are attached to the breech ring (fig. 28). The variable recoil control rod, which has been automatically adjusted to the elevation of the weapon, remains stationary in its bore in the recoil piston rod.

(2) The oil in the recoil cylinder in the path of the piston is forced through ports in the piston rod, through the grooves in the control rod,

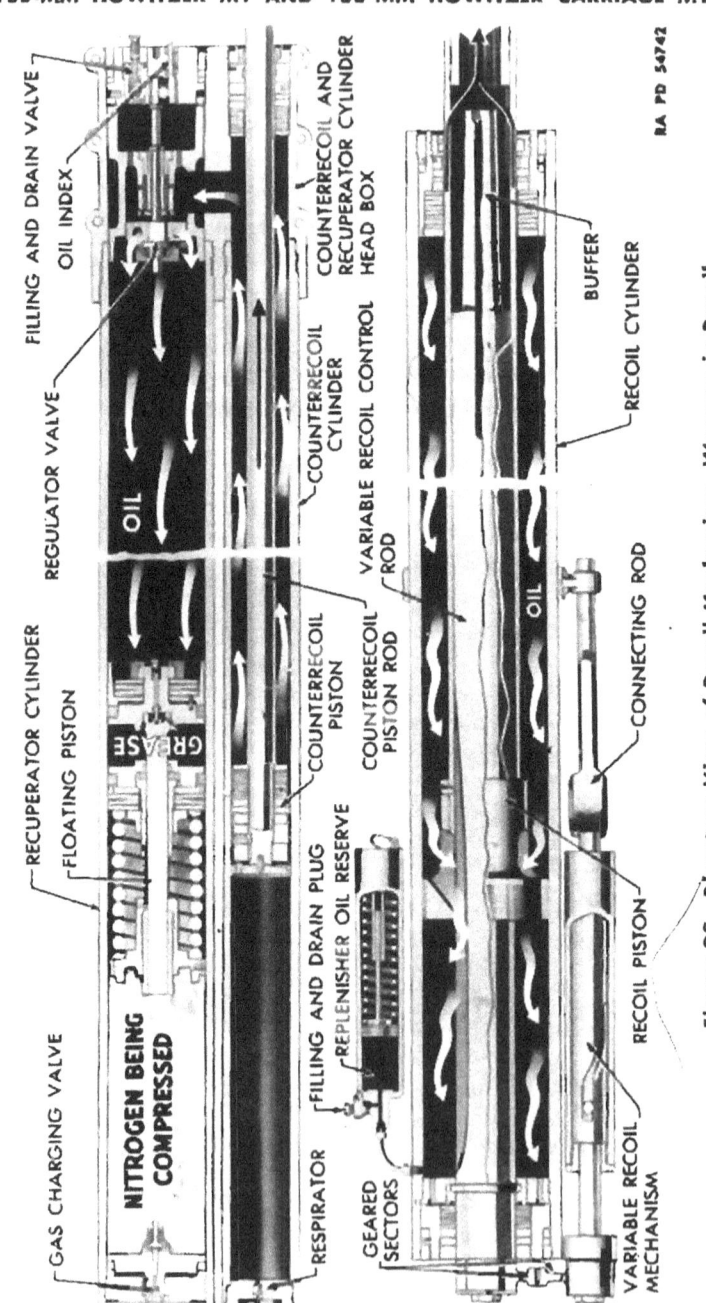

Figure 28 — Phantom View of Recoil Mechanism — Weapon in Recoil

DESCRIPTION AND FUNCTIONING OF CARRIAGE

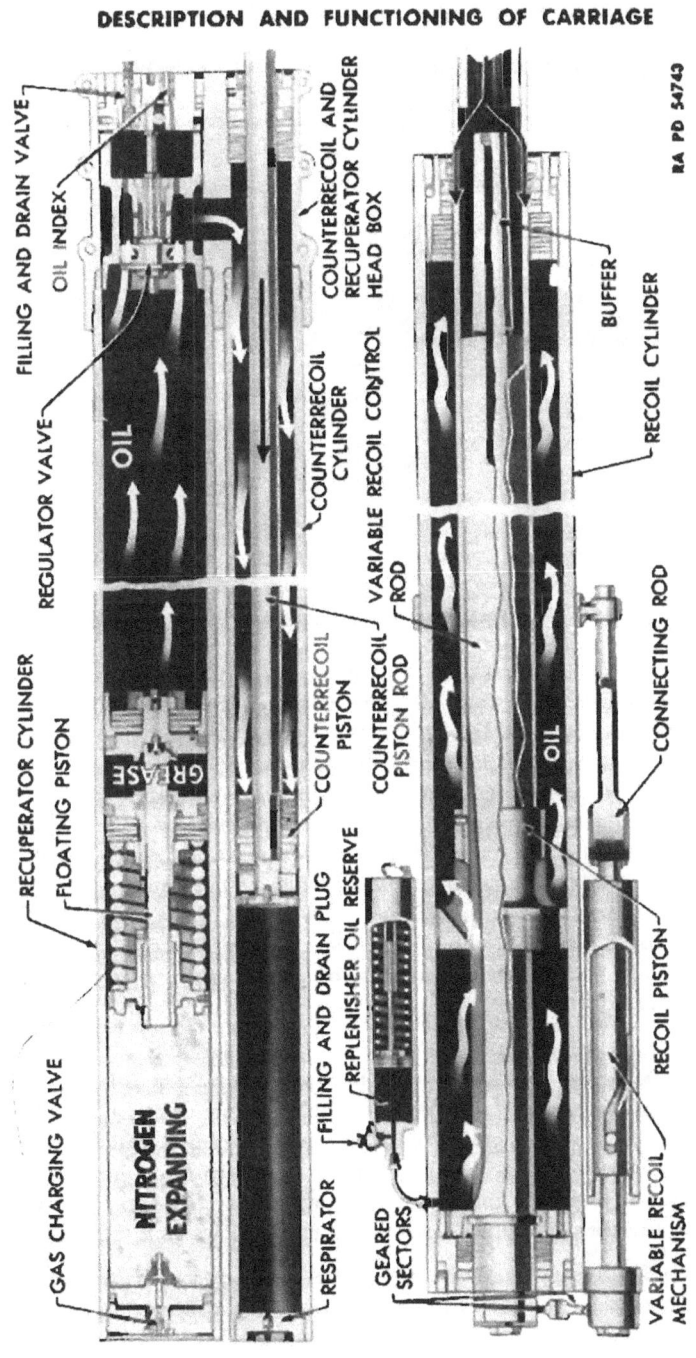

Figure 29—Phantom View of Recoil Mechanism—Weapon in Counterrecoil

155-MM HOWITZER M1 AND 155-MM HOWITZER CARRIAGE M1

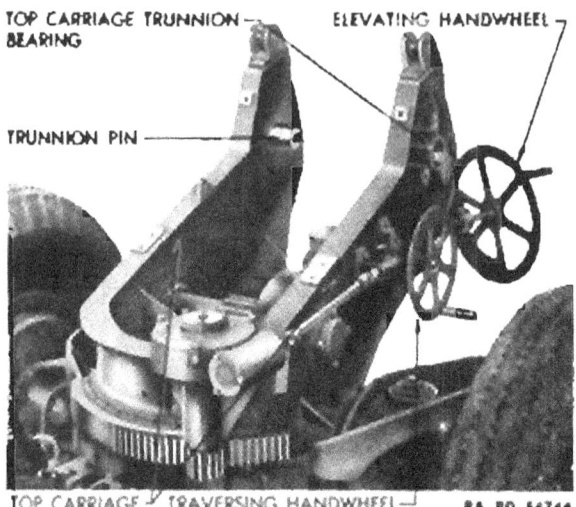

Figure 30 — Top Carriage, One Trunnion Pin in Position

to the other side of the piston. This has a throttling action on the oil, causing resistance which absorbs a great portion of the recoil energy.

(3) At the same time, the oil in the counterrecoil cylinder in the path of the counterrecoil piston is forced through the communicating opening in the walls of the counterrecoil and recuperator cylinders. This forces open the spring-loaded regulator valve, and the oil forces the floating piston forward, further compressing the nitrogen in the space ahead of the floating piston. The resistance which must be overcome to build up the nitrogen pressure counteracts the recoil action of the weapon.

(4) The throttling of the oil, the increased nitrogen pressure, and the combined friction of the packings of the recoiling parts bring the weapon to rest.

c. **Counterrecoil Action.** When the weapon has fully recoiled, the highly compressed nitrogen immediately begins to expand, forcing the floating piston and the oil in back of it in the opposite direction (fig. 29). The regulator valve is closed, and the oil is forced to return to the counterrecoil cylinder through the counterrecoil controlling holes in the regulator valve. In the counterrecoil cylinder, the oil forces the piston forward, and the counterrecoil piston rod pulls the weapon into battery.

d. **Buffer Action.** Near the end of the counterrecoil movement, the buffer chamber at the end of the recoil piston rod encloses the rear end of the variable recoil control rod (fig. 29). Oil, accumulated in the cham-

TM 9-331
13-14

DESCRIPTION AND FUNCTIONING OF CARRIAGE

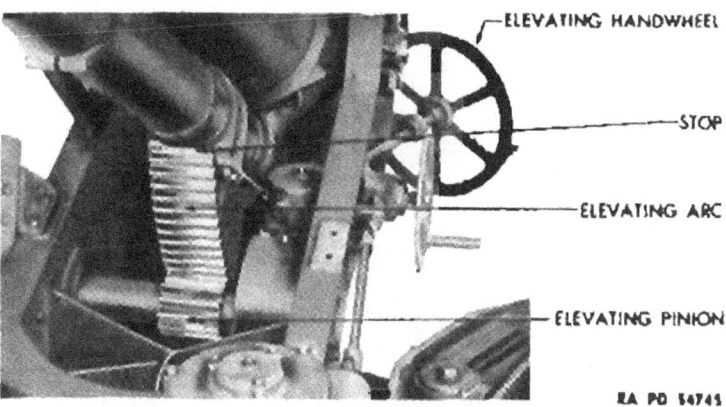

Figure 31—Elevating Mechanism

ber during recoil, is trapped. It is forced out by way of the throttling grooves in the buffer and an oil passage in the control rod. This throttling of the oil cushions the final movement of the weapon as it is drawn into battery.

14. TOP CARRIAGE.

a. The top carriage (fig. 30), of welded steel construction, has a circular base and 2 rearwardly extending arms. It supports the tipping parts of the weapon, the elevating and traversing mechanism, the shields, and the sighting equipment. It is supported by and rotates on the bottom carriage to which it is secured by the pintle (par. 18 d).

b. Connections for the rear ends of the equilibrators are provided on the upper ends of the arms. A connection for the variable recoil connecting rod is located on the upper edge of the right arm.

c. Trunnion Bearings. The howitzer, cradle, and recoil mechanism are supported on trunnion pins (fig. 30) which extend through the trunnion bearings in the side arms of the top carriage and into roller bearings mounted in the trunnion bearings of the cradle. The trunnion pins and the inner rings of the roller bearings, which are secured to the pins by retainers, remain stationary with the pins. The outer rings of the roller bearings are secured to the cradle by caps and tip with the cradle. Each assembly is oil-sealed.

d. Shields. The armor plate shields (fig. 38) are secured to the top carriage. The top portion of the left shield is hinged and may be secured in the desired position by the locking device. The sighting equipment case, mounted on the left shield, is equipped with a hinged cover and

TM 9-331
14-16

155-MM HOWITZER M1 AND 155-MM HOWITZER CARRIAGE M1

Figure 32 — Traversing Mechanism

padlock and is for stowage of the panoramic telescope when the latter is not in use.

15. ELEVATING MECHANISM.

a. The elevating mechanism (fig. 31) controls the movement of the weapon in elevation and depression by tipping the cradle and weapon on the trunnion pins secured in the arms of the top carriage. The elevating handwheel is mounted on a shaft which extends to the front through the left arm of the top carriage. Limited by stops at the ends of the elevating arc, which is mounted on the underside of the cradle, the weapon has a range in elevation of 1,156 mils (65 degrees).

b. Motion of the elevating handwheel is transmitted through shafts, a flexible joint, gears, a worm, and a worm wheel to the elevating pinion which meshes into the elevating arc. The pinion, worm wheel, worm, and handwheel shaft are mounted on antifriction bearings. An oiltight gear housing and oil retainers for the bearings provide means for constant lubrication.

16. TRAVERSING MECHANISM.

a. The traversing mechanism (fig. 32) rotates with the top carriage and controls the movement of the weapon in azimuth by means of the traversing pinion which is rolled in the teeth of the stationary traversing arc mounted on the left front side of the bottom carriage. The traversing handwheel is mounted on the left side of the left arm of the top carriage.

DESCRIPTION AND FUNCTIONING OF CARRIAGE

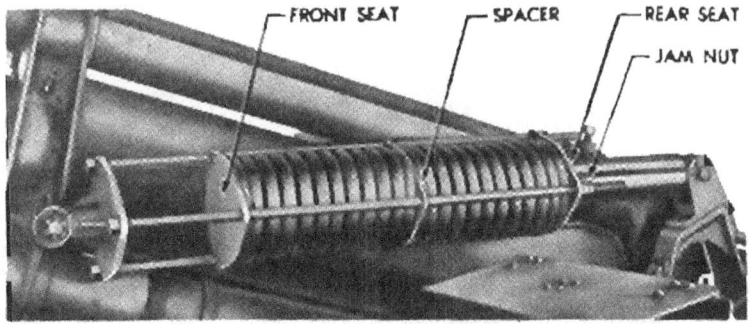

Figure 33—*Equilibrator and Connections to Top Carriage and Yoke*

Limited by stops on the bottom carriage, the weapon has a range in azimuth of 471 mils (26½ degrees) each side of mid-position.

b. Motion of the traversing handwheel is transmitted through gears, flexible joints, a shaft, a worm, and a worm wheel to the traversing pinion which meshes into the traversing arc. The worm and worm wheel are mounted on antifriction bearings. Oiltight housings and oil retainers, seals, and covers provide means for constant lubrication. The mechanism is mounted on the left to provide clearance for the weapon at zero elevation.

17. EQUILIBRATORS.

a. Two spring type equilibrators (fig. 33) are provided to neutralize the unbalanced weight of the weapon and to reduce the manual effort required to elevate it. The equilibrators connect the recoil yoke with the upper ends of the top carriage. The springs are compressed as the howitzer is depressed, counterbalancing the muzzle-heavy weight of the tipping parts and eliminating the need for manually braking their descent. The energy stored in the compressed springs is released as the howitzer is elevated; the springs are permitted to expand, applying lifting pressure to the weapon, and assisting in elevating it.

b. The 4 springs in each equilibrator are assembled in pairs, with 1 spring inside the other. They are assembled on a tube with the pairs of springs separated by a disk-like spacer which slides on the tube. They are retained at the front by a disk-like seat which is pinned to the front end of the equilibrator tube. A similar seat at the rear of the springs slides on the tube and compresses the springs through the action of 3 rods.

c. The front ends of the rods are connected by an end to the recoil yoke. The rear end of the tube is pinned through a roller bearing to the forked connection on an arm of the top carriage. The assembly is kept

155-MM HOWITZER M1 AND 155-MM HOWITZER CARRIAGE M1

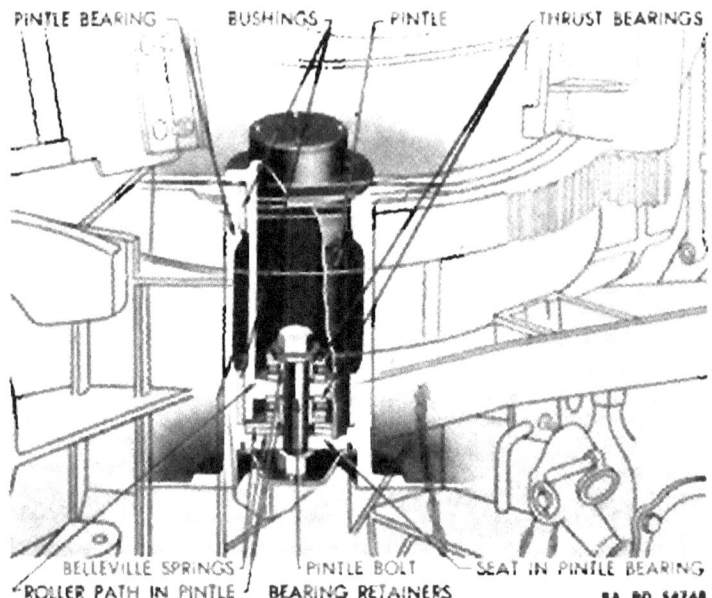

Figure 34 — Cutaway View of Top Carriage and Bottom Carriage Showing Pintle, Pintle Bearing, and Connection

in alinement by the front seat and the spacer sliding on the rods and by the rear seat and the spacer sliding on the tube. Adjustment of the equilibrators is by means of the nuts and jam nuts on the rear ends of the rods.

18. BOTTOM CARRIAGE.

a. The bottom carriage (fig. 34) is of welded steel construction. It supports the weight of the carriage and weapon on the axle and wheels. It transmits firing stresses to the firing jack and the trails. The firing jack, the traveling lock, and the traversing arc are mounted on the front of the bottom carriage. The sides of the bottom carriage are drilled to receive the trail hinge pins.

b. A liner of large diameter is mounted on the top surface of the bottom carriage. When the howitzer is fired, the smoothly finished undersurface of the top carriage bears on this liner. When the howitzer is traversed, the weight of the top carriage and the parts it supports are traversed on the pintle and the pintle bearing, and not on the bearing surfaces of the top carriage and the liner.

c. **Pintle Bearing.** The pintle bearing (fig. 34) is a vertical, cylindrical bore through the center of the bottom carriage. It is lined with

TM 9-331
18

DESCRIPTION AND FUNCTIONING OF CARRIAGE

Figure 35 — Firing Jack in Traveling Position

bushings and is provided with a seat near the bottom which closes the bore except for a bolthole through the center of the seat.

d. Pintle.

(1) The pintle (fig. 34) is a tubular projection extending from the bottom surface of the top carriage. It is the pivot for the movement of the weapon in traverse. It is shackled to the pintle bearing by an antifriction bolt and nut arrangement (fig. 34) which prevents lifting of the top carriage from the bottom carriage during firing or when the weapon and carriage are being drawn.

(2) The pintle rotates in the pintle bearing of the bottom carriage. It is borne on thrust bearings which contact the upper and lower surfaces of a roller path welded inside the pintle near its lower end. The lower thrust bearing rides on Belleville springs which rest on the seat near the bottom of the pintle bearing. The upper bearing is held against the roller path by a bolt which extends through and clamps the upper bearing, the roller path, the lower bearing, and the Belleville springs to the seat in the pintle bearing.

(3) The Belleville springs on the seat in the pintle bearing slightly raise the pintle and the top carriage off the liner for greater ease of traversing. The force of recoil compresses these springs, permitting contact

155-MM HOWITZER M1 AND 155-MM HOWITZER CARRIAGE M1

Figure 36—Firing Jack in Firing Position

of the bearing surfaces of the top carriage and liner. The pintle assembly is sealed above and below by covers.

19. **FIRING JACK.**

  a. The firing jack (figs. 35 and 36) is bolted to the front of the bottom carriage, to which it remains attached at all times. The weapon is fired with the carriage resting on the firing jack and the wheels raised clear of the ground.

  b. The toothed rack plunger of the firing jack is raised and lowered by ratchet mechanisms on the ends of a shaft. The ratchet mechanisms are provided with short, permanently-connected handles for quick-action operation of the plunger to loaded position, and long, removable handles to raise or lower the loaded plunger. The removable handles are carried on the front of the right shield.

  c. The jack float (fig. 36) is provided to give the lower end of the plunger sufficient bearing surface on soft ground. The float is carried on top of the trails in traveling position. In use, the flattened sides of the end of the plunger are inserted in the slotted seat in the center of the float. Then, the float is turned 90 degrees to lock it in place.

TM 9-331
19-20

DESCRIPTION AND FUNCTIONING OF CARRIAGE

RA PD 54711

Figure 37 — Traveling Lock in Locked Position

d. In firing position (fig. 36), the plunger is depressed until the jack key can be inserted through the openings in the jack housing to seat against the top of the plunger. When the plunger is raised against the jack key, the jack is locked and the firing stresses are taken off the moving parts of the jack mechanism and transmitted directly to the float.

e. In traveling position (fig. 35) the lower end of the rack plunger is protected from dirt and foreign matter by a small float and locked in position by a pin. The openings in the jack housing for the jack key are closed by covers. The ratchet wheels are engaged by spring-loaded plungers. The handles of the plungers are rotated 180 degrees to reverse the movement of the ratchets. The handles are inscribed with arrows which indicate the direction of movement for which the ratchets are set.

20. TRAVELING LOCK.

a. The traveling lock (fig. 37) is a triangularly shaped casting which is hinged to the front of the bottom carriage and is secured to the bottom of the recoil yoke by a lockpin when the materiel is in traveling position. In firing position, the lock is held horizontally by lugs on the lower portion of the lock contacting mating seats on the firing jack housing. The lockpin has a partial flange near its handle by means of which it is locked in position.

TM 9-331
20-21

155-MM HOWITZER M1 AND 155-MM HOWITZER CARRIAGE M1

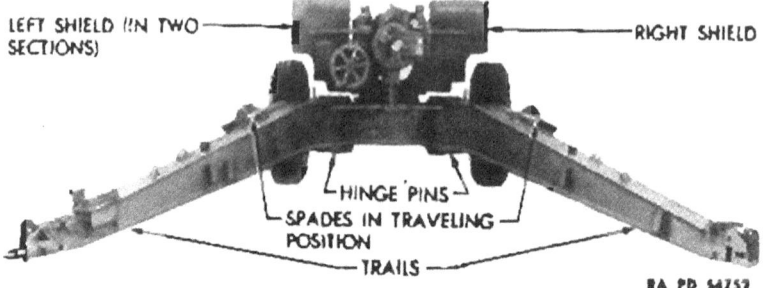

Figure 38—Carriage from the Rear with Trails Spread

21. **TRAILS.**

a. The 2 trails (fig. 38) are welded steel box type girders. They are hinged to the bottom carriage by the trail hinge pins. The ends of the top and bottom trail hinges are provided with shoulders which contact stops on the bottom carriage to control the maximum trail spread. When spread, each trail forms an angle of 30 degrees with the center of the carriage.

b. Stops, welded to the inner surfaces of the trails, contact the rear of the bottom carriage and control the closing of the trails to prepare for traveling position. When closed, the trails are secured by the trail lock and a projection on the left trail which seats in a mating hole in the right trail and is locked by a lock pin.

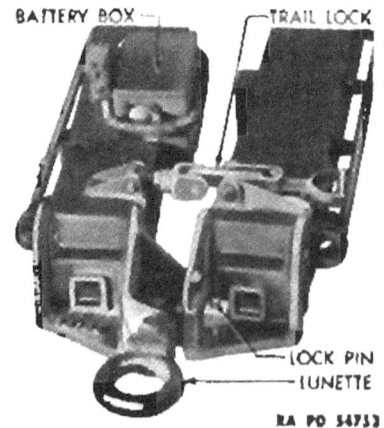

Figure 39—Trail Lock in Locked Position

TM 9-331
21

DESCRIPTION AND FUNCTIONING OF CARRIAGE

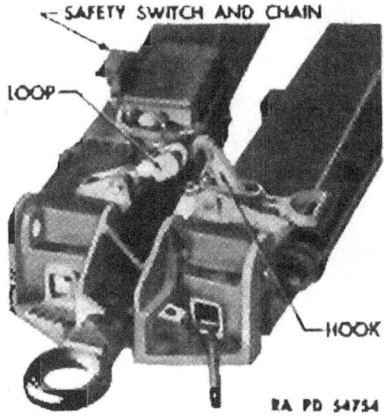

Figure 40 — Trail Lock in Unlocked Position

c. **Trail Lock.** The trail lock (figs. 39 and 40) is a toggle type clamping mechanism operated by a handle. The handle controls the action of a hook which engages a loop or stirrup-shaped fitting on the opposite trail. When the handle is lifted and swung toward the opposite trail, the hook is disengaged from the loop. The loop is threaded to permit the adjusting of the tension of the trail lock.

d. **Lunette, Battery Box, and Maneuvering Handles.** The lunette (fig. 39) is assembled in the end of the left trail by 2 pins. The battery box, located near the end of the left trail, houses dry cell batteries which supply current for operating the brakes through the safety switch. The safety switch (fig. 39) is mounted on the outside of the battery box.

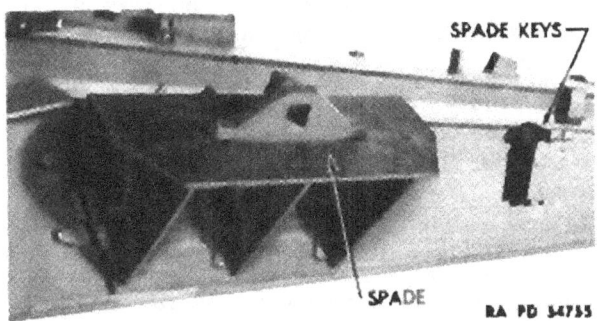

Figure 41 — Spade and Spade Keys in Travelling Position

155-MM HOWITZER M1 AND 155-MM HOWITZER CARRIAGE M1

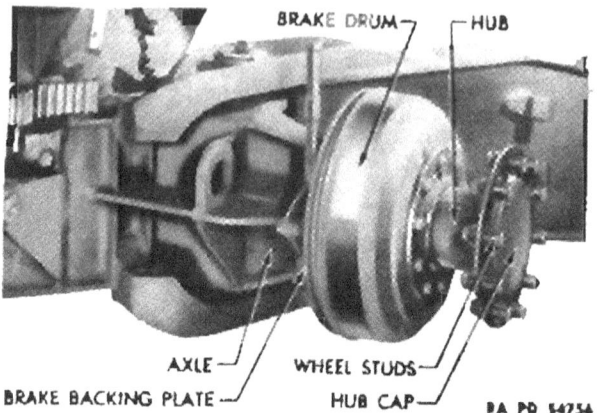

Figure 42 — Axle with Brake Drum and Hub

Tubular steel handles are provided on the outer rear ends of the trails for convenience in maneuvering the trails. In traveling position, the jack float is carried on the upper surfaces of the trails.

c. *Spades.* In firing position, the spades are mounted in the brackets on the rear ends of the trails; each spade is secured to its trail by a rectangular key, and the spades are buried in the ground. The spades are built of plates and function as spades and floats. In traveling position, the spades (fig. 41) and keys are carried in brackets on the outer sides of the trails.

22. AXLE AND HUBS.

a. *Axle.* The axle (fig. 42) consists of tubes with taper bores welded to the sides of the bottom carriage and reinforced by plates. The axle spindles are seated in the taper seats in the tubes and are secured in place by taper pins. The backing plates of the brakes are mounted on flanges on the ends of the tubes.

b. *Hubs.* The hubs (fig. 42) are supported on the bearing surfaces of the axle spindles by tapered roller bearings. The brake drums are bolted to the inner flanges of the hubs while the wheels are fastened to the outer flanges of the hubs by studs and nuts.

c. The inner bearing (fig. 43) of each wheel seats against a shoulder inside the hub. This bearing and its oil retainer are a drive fit in the hub, and both are secured in position by a retaining ring. This bearing is reached for greasing and inspection by removing the hub from the spindle, removing the brake armature, grease guard and bearing retaining ring, and driving the bearing and the oil retainer from the hub.

TM 9-331

DESCRIPTION AND FUNCTIONING OF CARRIAGE

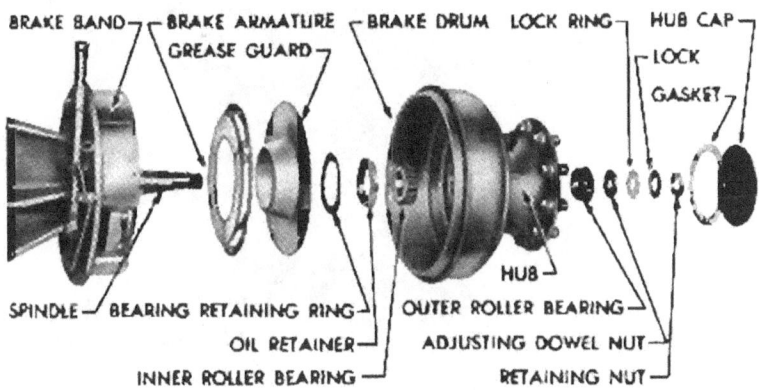

Figure 43 — Exploded View of Brake Drum and Hub Showing Bearings and Bearing Adjustment and Retaining Parts

d. The outer bearing (fig. 43) and the entire hub assembly are secured to the spindle and are adjusted by means of an adjusting dowel nut, a lock ring, a lock, and a hexagonal retaining nut. The adjusting dowel nut is screwed against the outer bearing until the bearings are properly adjusted. This nut is provided with a dowel which enters the appropriate hole of the lock ring, which is placed next to it on the spindle. The lock

Figure 44 — Wheel and Tire Showing Tire Retaining Ring

51

TM 9-331
22-23

155-MM HOWITZER M1 AND 155-MM HOWITZER CARRIAGE M1

Figure 45 — Brake Mechanism in Drum and on Backing Plate

ring, in addition to having a number of holes near its circumference, has a hole in its center shaped to fit the flattened upper surface of the spindle. The shape of this hole does not permit the lock ring to turn on the spindle.

e. The lock is placed on the spindle after the lock ring has been seated, and the hexagonal retaining nut is screwed tightly on the spindle. The lock is a soft metal disk, a portion of the circumference of which is bent against one of the edges of the hexagonal retaining nut to lock this nut in place. A hub cap, fastened over the outer end of the bore of the hub by 4 screws, protects the inside of hub against the entrance of dirt and other foreign matter.

23. WHEELS AND TIRES.

a. **Wheels.** The wheels (fig. 44) are dish-shaped disks, the flanges of which are riveted to the rims. The wheels are fastened to studs in the hubs by 10 nuts with right-hand threads when the tire is to be removed from a wheel. The tire must be deflated before removing the 18 nuts and the retaining ring and sliding the tire and tube off the rim.

b. **Tires.** The tires (fig. 44) are 14.00—20-inch, military combat, 16-ply, heavy-duty, nonskid, balloon tires. They are equipped with 14.00—20-inch (air container) bullet-resisting truck tubes. The recommended pressure for these tires is 65 pounds. NOTE: If the wheel is to be removed complete with tire, remove the inner row of 10 nuts only. It is not necessary to deflate the tire.

TM 9-331
24

DESCRIPTION AND FUNCTIONING OF CARRIAGE

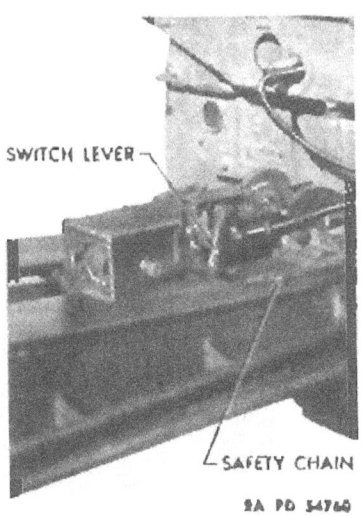

Figure 46 — Safety Switch for Emergency Brake Application

24. ELECTRIC BRAKES.

a. The power brakes (fig. 45) are of the electrically operated type. Each wheel unit is composed principally of a circular armature which is assembled to and revolves with the wheel, and a circular magnet and brake band which are mounted on the stationary backing plate of the brake mechanism. Brake application is controlled from the prime mover to which the howitzer carriage is limbered.

b. Electric current from the battery of the prime mover causes the magnet to cling to and revolve with the armature on the wheel until the lug on the magnet expands the brake band against the brake drum and stops the wheel. The greater the current, the tighter the magnet will cling to the armature.

c. When the current is shut off, the armature is released by the magnet and a spring is allowed to pull the brake band away from contact with the brake drum. A slight slipping action between the armature and magnet eliminates grabbing and locked brakes.

d. Safety Switch.

(1) The safety switch (fig. 46) automatically applies the brakes on the howitzer carriage wheels in the event of a break-in-two between

TM 9-331
24-25

155-MM HOWITZER M1 AND 155-MM HOWITZER CARRIAGE M1

Figure 47 — Hand Brake Mounted on Brake Backing Plate

the carriage and the prime mover. The safety switch is located on the battery box on the left trail.

(2) As the break-in-two occurs, the safety chain, which connects the safety switch with the prime mover, pulls the switch lever. This closes the circuit from the dry batteries in the battery box on the left trail to the brakes, energizes the magnet, and sets the brakes.

25. HAND BRAKES.

a. The hand brake mechanisms are operated by brake levers (fig. 47) assembled to the outer sides of the brake backing plates. Each lever is secured in the engaged position by a spring-loaded catch engaging the teeth of a rack. The knob at the top of the brake lever is pressed down to release the lever.

b. The lower end of the lever is mounted on one end of a shaft by a connection which permits adjustment of the position of the lever when the brake band has become worn. The other end of the shaft is assembled with a cammed roller (fig. 45) which contacts the thrust lever to force the brake band against the brake drum.

TM 9-331
26-28

Section III

OPERATION

	Paragraph
Introduction	26
To operate breech mechanism	27
To place weapon in firing position	28
To check liquid in recoil mechanism	29
To traverse	30
To elevate	31
Prior to firing	32
To load	33
To fire	34
Observation during firing	35
To unload	36
To remove a fuze from a shell	37
To place in traveling position	38
Brake operation	39

26. INTRODUCTION.

a. This section outlines the operation of the weapon and carriage. It prescribes precautions to be taken for the protection of the personnel and materiel. Many of the operations described herein will be performed simultaneously. This section is not to be construed as a field manual on the service of the piece.

27. TO OPERATE BREECH MECHANISM.

a. To Open Breech. Remove Firing Mechanism M1. With the left hand, release the operating lever latch, while with the right, pull the operating lever backward and downward until it contacts its stop. Then swing the operating lever in a horizontal arc to the right.

b. To Close Breech. Swing the operating lever in a horizontal arc to the left. When the breechblock rotating roller enters the breechblock rotating cam, raise the operating lever upward and forward until the operating lever latch latches the lever.

28. TO PLACE WEAPON IN FIRING POSITION.

a. Maneuver the weapon into position with the prime mover, with the muzzle pointing in the direction of fire. Set the hand brakes on the piece.

b. Disconnect the jumper cable and safety chain at the prime mover; wrap them around the battery box on the left trail. Unlimber the carriage by disconnecting the lunette from the pintle of the prime mover. Withdraw the prime mover from the vicinity of the piece.

REPLENISHER PISTON GUIDE

Figure 48 — Caging Oil Reserve in Replenisher

c. Remove the trail lockpin, release the trail lock, and spread the trails sufficiently to remove the firing jack float from its fastenings on the tops of the trails. Place the float in position under the firing jack.

d. Spread the trails to their fully open position against their stops on the bottom carriage. On hard ground, mark the positions the spades will occupy and dig pits for the spades; assemble the spades to the trails and lock with the spade lock pins. On soft ground, assemble the spades to the trails and dig in front of the spade blades to permit them to sink into the ground (the weight of the trails on the spades will facilitate this work). The horizontal floats of the spades should come in contact with the ground.

e. Disconnect the traveling lock. Remove the traveling lock pin from the bottom of the firing jack. With the quick-action handles, lower the rack plunger of the firing jack and remove the traveling float from the lower end of the rack plunger. Seat the lower end of the rack plunger in the float and lock the float in place by rotating it a quarter turn, or 90 degrees.

f. Remove the firing jack handles from the front of the right shield and insert them in the ratchets. Point the arrows on the ratchet handles down. Work the firing jack handles to raise the carriage until the jack key can be inserted from the right through the openings in the sides of the jack over the top of the rack plunger. Then point the arrows on the ratchet handles up and lower the carriage (raise the plunger) until the top of the plunger contacts the jack key. This will take the weight of the weapon and carriage off the firing jack mechanism. The wheels will be clear of the ground.

OPERATION

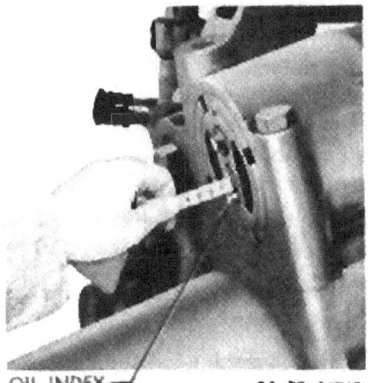

Figure 49 — Measuring Recuperator Oil Index

29. TO CHECK LIQUID IN RECOIL MECHANISM.

a. Before firing, the recoil mechanism should be checked to determine if there is the proper amount of liquid in the replenisher and the recuperator. This check also should be made after every tenth round. Insufficient liquid (oil reserve) in the recoil mechanism when the howitzer is fired will result in damage to the mechanism.

b. **To Gage Oil Reserve in Replenisher.**

(1) The position of the replenisher piston indicates the amount of oil reserve in the replenisher. To measure the position of the replenisher piston, pull out the plug from the center of the piston guide at the rear end of the replenisher, insert a scale in the opening, and push it in until it comes in contact with the rear end of the piston extension (fig. 48). Read on the scale the graduation that is flush with the rear face of the replenisher.

(2) The normal position of the replenisher piston extension is 5½ inches (140-mm) from the rear face of the replenisher. This position indicates a full recoil cylinder and a proper oil reserve in the replenisher.

(3) When the end of the piston is within 3½ inches (89-mm) of the rear end of the replenisher, it indicates too much reserve oil. Oil should be removed from the replenisher before firing is continued. This is done by means of the oil release.

(4) When the end of the piston is 7½ inches (190-mm) or more from the rear end of the replenisher, it indicates no reserve, and sufficient oil should be added to establish the proper reserve. To establish the replenisher oil reserve, see paragraph 51 d.

155-MM HOWITZER M1 AND 155-MM HOWITZER CARRIAGE M1

c. *To Gage Oil Reserve in Recuperator.*

(1) The position of the oil index indicates the oil reserve in the recuperator. The oil index is located in the recuperator rear cylinder head below the filling and drain valve. To measure the position of the oil index, hold one end of a scale firmly against the cylinder head with the edge of the scale parallel to and close to the oil index (fig. 49). Read on the scale the graduation opposite the rear end of the oil index.

(2) The normal position of the oil index is ¼ inch (6-mm) out from the rear face of the cylinder head. To establish the recuperator oil reserve, see paragraph 51 d.

30. TO TRAVERSE.

a. The traversing handwheel is located on the left side of the carriage ahead of the elevating handwheel. One complete turn of the handwheel in a clockwise direction traverses the carriage to the right 9.6 mils (32 degrees, 4 minutes). The range of traverse is 942 mils (53 degrees), or 471 mils (26½ degrees) each side of mid-position. Approximately 84 turns of the handwheel are required to traverse the mount from one side to the other of its range of traverse.

31. TO ELEVATE.

a. The elevating handwheel is located at the left rear of the carriage in back of the traversing handwheel. One complete turn of the handwheel in a clockwise direction elevates the weapon 14.8 mils (50 minutes). The range of elevation is 1,156 mils (65 degrees). Approximately 78 turns of the handwheel are required to elevate the weapon from 0 to 65 degrees.

32. PRIOR TO FIRING.

a. Remove the telescope from its case on the rear of the left shield and place it in position in its mount. Assemble the firing accessories.

b. Check the replenisher and recuperator oil reserves (par. 29). When it is known that rapid fire is to take place, release oil from the replenisher until the rear end of the piston is 7½ inches (190-mm) from the rear of the replenisher. The heat of firing should quickly bring the piston to its normal position due to expansion of the oil.

c. Inspect the bore and breech to see that no dirt or foreign matter has accumulated. If there is foreign matter, clean thoroughly and wipe dry. Normally, the bore should be wiped or cleaned to remove the coating applied after the previous firing. Examine the breechblock, primer vent, and gas check pad. Inspect the recoil mechanism for oil leakage. Make certain that the recoil slide (the exposed bearing surface of the weapon) is clean and well lubricated.

TM 9-331
33

OPERATION

33. TO LOAD.

a. Lower the weapon to an elevation suitable for loading (about 150 mils). Remove the Firing Mechanism M1 and open the breech. Latch the percussion hammer in the released position by means of the percussion hammer latch.

CAUTION: The percussion hammer latch pin will not be drawn from its latched position, where it interferes with the upward swing of the percussion hammer, until after the breech has been closed and locked and the piece is ready to be fired. This is a safety precaution.

b. Swab the powder chamber and breech recess. If a charge has been fired, wipe off the powder residue from the obturator spindle head, gas check pad, gas check seat, and the threaded sectors of the breech recess and breechblock with a cloth or piece of waste slightly dampened with oil. Clean the primer vent with the vent cleaning bit. Inspect the bore for burning fragments of powder bags or other objects, and for bore injuries. In night firing, swab the bore with water.

c. Prepare the Projectile. Verify the type, weight, and lot number, and examine carefully for defects. Remove the grommet and inspect the rotating band with special care; remove any burs with a file. Clean the entire surface of the projectile with a piece of waste, or with a sponge and water. Sand or dirt on the projectile might cause premature detonation when the piece is fired and will cause undue wear on the bore.

d. Fuze the Projectile. Unscrew the eyebolt lifting plug from the fuze socket. Insert the designated fuze, being careful that it is fitted with its felt or rubber washer. Screw it home by hand. Give the fuze its final seating with the Fuze Wrench M7. No great force should be used. If there is any difficulty in screwing the fuze home, the fuze should be removed and another inserted. If the same trouble is encountered with the second fuze, the shell should be rejected. Set the fuze.

e. Bring up the projectile on the loading tray. Grasp the handles of the tray and raise it with the front slightly above the rear. Get a firm grip on the handles as a shell may be dropped easily if the tray is not carried in the proper position. Exercise extreme care that the fuze does not come in contact with anything.

CAUTION: The projectile will not be brought to the rear of the weapon until after the weapon has returned to battery.

f. Place the lip of the tray in the breech recess. Place the rammer head squarely against the base of the projectile, pushing it slowly until it has cleared the threads of the breech recess. Then ram the projectile home with a powerful stroke. It is important that projectiles be rammed home with a uniform force since variations in the ramming force will cause slight variations in the range. It is also important to avoid damage to the rotating bands on the projectile as such damage is likely to cause erratic flight of the projectile.

g. **Prepare the Propelling Charge.** For description and preparation for firing of the propelling charge, see paragraph 95 e. Bring the prepared propelling charge up to the breech immediately after the projectile has been rammed.

CAUTION: An exposed propelling charge will not be near the weapon at any other time.

h. **Load the Propelling Charge.** The loading tray is not required. Place the charge in the chamber with the igniter end to the rear and push it in until the base of the charge is flush with the rear end of the chamber. The igniter pad must come directly in front of the vent when the breech is closed to insure ignition of the charge.

i. **Close the Breech.** To insure transmission of the flash from the primer to the charge the obturator spindle head must come in contact with the base of the charge when the breech is closed, must push the charge forward to its final position, and remain in contact with it.

34. TO FIRE.

a. Insert a percussion type primer in the Firing Mechanism M1. The primer case is inserted into the primer holder by pressing the head of the case downward firmly against the firing pin guide so that the rim of the case slides under the edges of the slot in the primer holder. The primer is then held in position by the pressure of the firing spring.

b. Should the primer be slightly oversize, or the primer holder dirty, the primer will stick before it is properly seated. Force should not be exerted. Remove the primer and clean the primer holder, or insert another primer.

c. Insert the firing mechanism in the firing mechanism housing, taking care that the front end of the primer has entered the obturator spindle plug. Seat the mechanism by turning the firing mechanism handle in a clockwise direction until it has engaged the latch. If the mechanism will not seat properly, the primer may be oversize, or its seat in the plug or in the primer holder may be dirty, or the breech may not be fully closed.

CAUTION: Make certain that the Firing Mechanism M1 is screwed home and is latched in position. Despite safety devices, it is possible to fire the piece even though the mechanism is not completely in its proper firing position. If this occurs, damage to the breech and injury to the personnel may result.

TM 9-331
34-35

OPERATION

d. Draw the percussion hammer latch pin to the left and out of the path of the hammer, turning the latch pin knob to place the pin in locked-out position. Attach the lanyard.

e. Grasp the handle of the lanyard with the right hand and without raising the hand, pull with a quick, strong pull (not a jerk) prolonged sufficiently to insure the percussion hammer hitting the firing pin. The lanyard will be pulled from a position as near the rear of the piece as is convenient, and sufficiently out of the line of recoil to insure safety. Care must be exercised in pulling the lanyard not to injure the firing mechanism parts.

f. If the long lanyard is used, it will be attached immediately before, and detached immediately after, the round is fired.

35. OBSERVATION DURING FIRING.

a. Observe the movement of the weapon in recoil. It should be smooth with uniformly decreasing velocity. The point of maximum recoil should be reached without shock. Then the counterrecoil system should return the weapon completely into battery without shock. If uneven, jerky movement or shock is observed, or if the weapon will not return fully to battery, inspect the recoil mechanism to determine the cause (par. 42 i through n).

b. Check the recoil mechanism to determine if there is any oil leakage. Measure the oil reserves in the replenisher and the recuperator at intervals during firing. The temperature of recoil oil rises during firing and causes the recoil oil to expand.

NOTE: In an emergency, when it is necessary to continue firing without interruption, firing may be permitted until the end of the replenisher piston is 2 inches (50-mm) from the rear face of the replenisher.

c. Measure the length of recoil for the first round and at intervals during firing, when practicable. The normal length of recoil of the 155-mm howitzer is 60 inches at 0- to 25-degree elevation and 40 inches at 40- to 65-degree elevation. Excessive recoil will cause damage to the mechanism.

d. If the length of recoil does not fall within the limits designated when the weapon is operating at normal temperature, check the oil reserves in the replenisher and recuperator immediately. If the oil reserves are abnormal, consult paragraph 51 and take the necessary corrective measures.

e. To Measure Length of Recoil. Place a heavy smear of grease on the recoil slide or bearing surface of the weapon, extending from the wiper on the front end of the yoke to the front end of the recoil slide

155-MM HOWITZER M1 AND 155-MM HOWITZER CARRIAGE M1

(or tie a piece of string around the recoil slide in front of the wiper). After the weapon has fired and returned to battery, measure the distance between the wiper and the point to which the wiper has moved the grease (or string).

36. TO UNLOAD.

a. **Service Rounds.** No unloading rammer is provided for use with this materiel for unloading service rounds of ammunition. When it is desired to unload the piece, the projectile may be fired out of the weapon.

b. **Dummy Projectile.** To unload the dummy projectile, lower the weapon to a convenient elevation (about 150 mils), and with the loading tray in place, remove the projectile with the dummy projectile extractor. Place the hook of the extractor in the recess in the base of the dummy projectile and engage the hook on the shoulder. Then jerk the projectile to release the band stuck in the forcing cone. Push the projectile forward and repeat if necessary. Use the extractor to guide the projectile onto the loading tray.

37. TO REMOVE A FUZE FROM A SHELL.

a. If, for any reason, a projectile which has been fuzed is not to be fired, the fuze will be removed. Reset the fuze to "SAFE" if so designed. Start the unscrewing operation with the Fuze Wrench M7; complete the unscrewing of the fuze by hand.

CAUTION: If the adapter starts to unscrew with the fuze, the unscrewing must be stopped at once and the shell disposed of as directed by the officer in charge.

38. TO PLACE IN TRAVELING POSITION.

a. If weapon has been fired, clean, dry, and oil the bore and chamber, and disassemble, clean, and oil the breech and firing mechanism as prescribed in paragraph 49.

b. Raise the carriage by means of the firing jack, remove the jack key, and lower the carriage until it rests on its wheels. Remove the firing jack float and replace the traveling float. Place the firing jack in traveling position. Connect the traveling lock to the recoil yoke. Secure the firing jack handles.

c. Remove the spade lockpins and close the trails, placing the firing jack float in its fastenings on the tops of the trails. Pry the spades out of the ground with the handspikes. Place the spades and spade lockpins in their carrying racks on the trails. Clamp the trails together with the trail lock and lock with the lockpin.

OPERATION

d. Release the hand brakes on the carriage. Back up the prime mover until the lunette can be placed over the pintle of the prime mover. Connect the jumper cable and safety chain.

39. BRAKE OPERATION.

a. The proper functioning of the brake control system is of vital importance. Wheels should be free from brake "drag," and the wheel bearings should be accurately adjusted to prevent drag due to loose bearings.

b. When applying the brakes for an ordinary stop, the handle of the controller on the prime mover should be advanced gradually for a light brake operation. Heavy brake application should be reserved for emergency stops and should not be employed in ordinary brake service. The load control on the controller is provided to permit the driver to regulate the braking power and prevent skidding regardless of load and road conditions. For slippery roads, the control should be set on "LIGHT."

CAUTION: To avoid injury to the personnel, to insure safe road transportation, and to prevent "jack knifing" of the load, the driver should have the load under control at all times by avoiding any slack between the load and the prime mover. On down grades, curves, and rough or slippery roads, the speed should be held to approximately 10 miles per hour. When applying the brakes for a slowdown or stop, always apply the brakes on the load before applying the brakes on the prime mover.

TM 9-331
40

155-MM HOWITZER M1 AND 155-MM HOWITZER CARRIAGE M1

Section IV

MALFUNCTIONS AND CORRECTIONS

	Paragraph
Misfire	40
Other malfunction of weapon	41
Malfunction of carriage	42
Malfunction of electric brakes	43

40. MISFIRE.

   a. A misfire occurs if the piece fails to fire when desired. Failure of the piece to fire is due to one of two causes: failure of the primer to fire or failure of the propelling charge to ignite.

   b. **General Precautions.** The following general precautions will be taken in all cases:

   (1) The piece will be kept trained on the target or on a safe place in the field of fire.

   (2) All persons will be kept clear of the path of recoil until after the breech is opened.

   (3) When ejecting the primer, opening the breech, or reaming the vent, the operator will stand clear of the path of recoil, and all other persons will be kept from the rear of the breech.

   (4) In no case will the breech be opened before the primer is removed.

   (5) Whenever a new primer is inserted and another attempt to fire results in failure, all precautions and procedure will be as prescribed for the first failure. The firing of more than two primers in an attempt to ignite the propelling charge usually is not justified.

   c. **Primer Failure (Discharge of Primer Is Not Heard After at Least Three Attempts Have Been Made to Fire the Primer, the Lanyard Being Pulled with Considerable Snap).**

Probable Cause	Probable Remedy
Defective primer. Improperly seated mechanism. Lack of, or insufficient, blow on primer.	If primer net is available, wait 2 minutes. If net is not available, wait 10 minutes. Remove firing mechanism, noting whether mechanism has been screwed home fully. If it has not been seated, this can have caused failure. Inspect primer. If head has been properly indented,

64

MALFUNCTIONS AND CORRECTIONS

Probable Cause — **Probable Remedy**

primer will be handled carefully and disposed of quickly because of possibility of hangfire. A new primer should be inserted and another attempt be made to fire.

Defective firing mechanism. Fouled firing pin; heavy or gummed oil; deformed point of firing pin; broken firing pin spring. — If head has not been properly struck, a new primer should be inserted in spare firing mechanism and an attempt be made to fire. First firing mechanism should be inspected for: dirty or gummy parts; lubricant too heavy for prevailing temperature; firing pin or firing pin spring broken; firing pin housing or primer holder loosened. Then wash parts with SOLVENT, dry-cleaning. Replace broken or damaged parts.

Hangfire. Damp or fouled vent through obturator; abnormal condition of propelling charge. — If primer has fired, a cleaning bit will be run through the vent, another primer will be inserted, and another attempt will be made to fire. If this attempt is unsuccessful, it may be assumed that the failure is due to the propelling charge.

d. **Propelling Charge Failure (Discharge of Primer is Heard, or Hangfire is Indicated).**

Abnormal condition of charge. Missing igniter; wet igniter; igniter charge folded over and not accessible to flash of primer; failure to remove igniter protector cap; increment section between base section and breech. — No attempt will be made to remove primer or open breech until 10 minutes have elapsed after firing of primer. After 10 minutes, firing mechanism will be removed, cleaning bit run through vent, a new primer inserted, and another attempt made to fire. If this attempt is unsuccessful, after a wait of 10 minutes, the breech will be opened, propelling charge re-

TM 9-331
40-41

155-MM HOWITZER M1 AND 155-MM HOWITZER CARRIAGE M1

Probable Cause	Probable Remedy
	moved and disposed of as directed by executive officer, and a new propelling charge inserted in the weapon.

41. OTHER MALFUNCTION OF WEAPON.

a. Breechblock Failure.

Seized breechblock.	Notify ordnance maintenance personnel.

b. Breech Mechanism Does Not Operate Freely.

Lack of lubrication, or scores on threads of breech recess or breechblock.	Disassemble breechblock and clean thoroughly. If threads are scored, repair must be made by ordnance maintenance personnel.

c. Breech Will Not Open or Fully Close with Firing Mechanism in Place.

Safety feature safety latch is functioning.	Remove firing mechanism. If mechanism is removed, latch may be stuck at right; move latch to left.

d. Threaded Sectors of Breech Ring and Breechblock Do Not Mate.

Improper assembly of breech mechanism; missing control arc; broken parts.	Insert missing control arc. Disassemble breech mechanism and reassemble in proper manner. Replace any broken or deformed parts.

e. Escape of Gases; Powder Fouling on Threads of Breechblock.

Bruised or burned gas check pad; burred or ruptured split rings; broken or weakened obturator spindle spring.	Disassemble breech mechanism; replace damaged obturator parts.

f. Operating Lever Does Not Latch Properly.

Weak or broken operating lever latch spring.	Disassemble operating lever latch and replace spring.

g. Percussion Hammer Not Working Freely.

Lack of lubrication; roughness of shaft or bearings.	Lubricate. If this does not correct malfunction, notify ordnance maintenance personnel.

MALFUNCTIONS AND CORRECTIONS

42. MALFUNCTION OF CARRIAGE.

a. Air in Replenisher or Recoil System.

Probable Cause

Replenisher piston packing defective; imperfect replenisher connections; leaking recoil cylinder.

Probable Remedy

Check replenisher connections; keep them tight and free from dirt. If air accumulates after replenisher has been drained and properly filled, refer to ordnance maintenance personnel.

b. Oil Leaks from Rear of Replenisher.

Whether or not a serious leak exists must be determined by frequency of filling required.

Oil may drip rapidly, or may run in a stream, from rear of replenisher when weapon is elevated, provided weapon has been at zero elevation for some time. A leak at any packing that does not exceed 3 drops per minute is not considered serious. If leakage is greater, notify ordnance maintenance personnel.

c. Position of Replenisher Piston Does Not Change During Firing.

Replenisher piston stuck.

Insert a block of hardwood in the rear of replenisher guide and against the end of the piston; tap block with hammer. Never strike walls of replenisher. If replenisher housing becomes dented in any way, notify ordnance maintenance personnel.

d. Oil Index Projects Less Than ¼ Inch (6-mm) When or Shortly After Oil Reserve Has Been Established.

Loss of gas pressure either through recuperator front cylinder head or past floating piston.

Gas escaping by floating piston is indicated by an emulsified condition of the reserve oil drained off. If when proceeding to establish oil reserve in ordinary manner, oil index does not move out and pump works easily, gas pressure has probably been lost. Substantiate this by attempting to drain reserve; oil will not spurt from a mech-

155-MM HOWITZER M1 AND 155-MM HOWITZER CARRIAGE M1

Probable Cause	Probable Remedy
	anism without at least some pressure. Notify ordnance maintenance personnel.

e. Oil Index Remains Stationary When Reserve Oil is Pumped in Against Evident Pressure.

Probable Cause	Probable Remedy
Packing too tight; index broken or locked by some foreign substance.	Drain off all reserve oil and refill. While injecting oil, tap oil index gently with each stroke of pump or each turn of oil screw filler. If oil index fails to move after 67 full strokes of Oil Pump M3, refer matter to ordnance personnel.

f. Excessive Leaks from Recuperator Filling and Drain Valve.

Probable Cause	Probable Remedy
Sticking of valve or defective packing.	Notify ordnance maintenance personnel.

g. Oil Drips from Counterrecoil Rod, Recoil Rod, or Control Rod Stuffing Boxes in Excess of 3 Drops Per Minute.

Probable Cause	Probable Remedy
Broken springs; more compression required on springs; damaged packing.	Notify ordnance maintenance personnel.

h. Oil Leaks from Forward End of Counterrecoil Cylinder.

Probable Cause	Probable Remedy
Black oil appearing in front of counterpiston is a normal condition due to lubrication. Clear oil is an indication of a leak due to broken packing springs or lack of compression on springs.	Report a leak of clear oil to ordnance maintenance personnel.

i. Weapon Will Not Return to Battery.

Probable Cause	Probable Remedy
Too much oil in replenisher.	Reduce amount of oil in replenisher to normal.
Insufficient oil in recuperator.	Drain off reserve oil and refill.
Insufficient gas pressure.	Report to ordnance maintenance personnel.

j. Weapon Returns to Battery with Too Much Shock.

Probable Cause	Probable Remedy
Insufficient oil in replenisher.	Refill replenisher to normal.
Excess oil in recuperator.	Drain recuperator reserve oil and refill.
Change of viscosity of oil due to heat of rapid firing.	Allow weapon to cool.

MALFUNCTIONS AND CORRECTIONS

Probable Cause	Probable Remedy
Friction factors of various packings are too low.	Report to ordnance maintenance personnel.

k. Weapon Slow to Return to Battery When Oil Indication is Normal.

Insufficient gas pressure; packing exerts too much friction.	Report to ordnance maintenance personnel.

l. Uneven or Jerky Counterrecoil.

Bearing surfaces fit too tightly, are scored, or lack proper lubrication; foreign substances in oil.	Report to ordnance maintenance personnel.

m. Weapon Recoils More Than Normal Allowable Distance.

Insufficient oil in recoil mechanism.	Refill replenisher to normal.
Insufficient gas pressure in recuperator; insufficient friction; malfunction of variable recoil mechanism.	Report to ordnance maintenance personnel.

n. Weapon Does Not Recoil Sufficiently.

High viscosity of oil due to low temperature.	After firing two or more rounds, the recoil will become normal if this is the cause.
Bearing surfaces fit too tightly, are scored, or lack lubrication; packing exerts too much friction; malfunction of variable control mechanism.	Report to ordnance maintenance personnel.

o. Weapon in Counterrecoil Does Not Cause Hissing Sound of Escaping Air.

Air vents are stopped up.	Clean the vents in the replenisher, which are continuations of test holes in piston guide. If air is not escaping from respirator in front end of counterrecoil cylinder, report to ordnance maintenance personnel.

p. Weapon Will Not Attain Maximum Elevation.

Variable recoil mechanism jammed; interference between tipping and non-tipping parts; malfunction of elevating mechanism.	Report to ordnance maintenance personnel.

TM 9-331
43

155-MM HOWITZER M1 AND 155-MM HOWITZER CARRIAGE M1

43. MALFUNCTION OF ELECTRIC BRAKES.

a. No Brakes or Intermittent Brakes.

Probable Cause	Probable Remedy
Broken wire in circuit.	Check entire wiring for broken wires.
Controller defective.	Short out controller by connecting both wires to one terminal and see if brakes are effective.
Poor connections.	Check, clean, and tighten all connections at brake, controller, load control, and socket.
Broken wire on magnet.	If broken wire is on outside of magnet, repair if possible. If no current flows through magnet, report to ordnance maintenance personnel.
Poor ground condition in circuit.	Clean up and tighten ground connections.
Defective plug or socket.	Check plug and socket for loose connections, dirty or corroded blades, or a broken socket. Repair or replace.

b. Weak Brakes.

Probable Cause	Probable Remedy
Worn out or greasy brake lining.	Lining may be worn to full extent of magnet travel. Refer to ordnance maintenance personnel for new lining.
Glazed or greasy magnet facing.	Roughen facing of magnet with coarse emery cloth.
Stop lights connected in circuit.	Check to determine if stop lights have been connected in circuit by mistake.
Wire broken in insulation. Loose connection. Poor contact at load control.	Check wiring for defective parts or connections. Short out load control.
Insufficient current.	Test with ammeter (par. 52 c (3)). Insufficient current may be caused by poor connections at brake, controller and load control, ground, or plug and socket. Clean up and tighten all connections, check plug and socket for corroded or dirty blades. Repair or replace socket.

70

TM 9-331
43

MALFUNCTIONS AND CORRECTIONS

Probable Cause	Probable Remedy
Poor ground connection at brakes.	Ground contacts must be solid and clean.
Worn wheel bearings; incorrect armature spacing.	Refer to ordnance maintenance personnel.

c. Brakes Grabbing.

Probable Cause	Probable Remedy
Only one brake working.	Test with ammeter (par. 52 c (3)).
Loose or worn wheel bearings.	Tighten bearings. Report to ordnance maintenance personnel for replacement.
Stop lights in brake circuit.	Determine if stop lights have been connected improperly.
Worn or greasy lining; drums out of round; bands distorted; lining loose on rivets; broken or weak band or magnet springs; controller burned out; poor contactor blade spacing; magnet bushing worn.	Refer to ordnance maintenance personnel.
Contactor arm in controller pitted.	Smooth out contactor arm with fine emery cloth.
Poor electrical connections.	Check wiring for loose connections and broken wires in insulation.

71

TM 9-331
44

155-MM HOWITZER M1 AND 155-MM HOWITZER CARRIAGE M1

Section V

LUBRICATION

	Paragraph
Introduction	44
Lubrication guide	45
Reports and records	46

44. INTRODUCTION.

a. General.

(1) Lubrication is an essential part of preventive maintenance, determining to a great extent the serviceability of parts and assemblies. Satisfactory operation and long life of the materiel are not assured unless the materiel is kept clean and well lubricated.

(2) Apply sufficient lubricants, but avoid wasteful practices. Excessive lubrication will result in dust accumulations on some moving parts and, if not removed, may cause wear and malfunctioning. Particular attention should be given to the lubrication of sliding surfaces of the breech mechanism and to other bearing surfaces which contain no oilholes, plugs, or fittings.

(3) Operate greasing and oiling devices slowly, maneuvering the parts while they are being oiled and greased to insure proper distribution of the lubricant to the bearing surfaces. Keep all exposed parts clean and well lubricated. The materiel should always be lubricated after washing. Do not use a high-pressure washing system for artillery materiel.

(4) Should an oiler valve stick and prevent the passage of oil, loosen it with a piece of wire pushed through the hole. Do not use a piece of wood; it might splinter. Care should be taken not to damage the valve. In cleaning oil fittings, the necessary wiping should be done with a piece of firm cloth and no lint should be allowed to remain in any opening.

b. Identification of Lubrication Points. Lubrication fittings are painted red for ease in locating. Oilholes are encircled by a red ring. In painting oil fittings, care must be taken to prevent paint from entering the fitting or to be placed where it will block the entrance of lubricant or be forced into the fitting.

c. Cleanliness. Care must be taken when cleaning any oil compartment or bearing surface to insure the complete removal of all residue or sediment, and to prevent dirt or other foreign matter from entering. Make sure all oilholes and fittings are clean before applying lubricant.

LUBRICATION

d. Climate. Operation under special climatic and atmospheric conditions, such as the extreme heat and humidity of the tropics, or Arctic cold, make necessary special treatment of the materiel. This special treatment may vary with the prevailing conditions and the seasons, and is to be applied at the discretion of the ordnance officer. Before lubricating materiel for Arctic conditions, thoroughly remove all old lubricant. It is not necessary to clean out old lubricant when materiel prepared for Arctic conditions is to be lubricated for warmer climates (chap. 7).

e. Cold Weather. For lubrication and service below 0 F, refer to OFSB 6-5.

45. LUBRICATION GUIDE.

a. General. Lubrication instructions for this materiel are consolidated in the lubrication guides (figs. 50 and 51). These specify the points to be lubricated, the periods of lubrication, and the lubricant to be used. In addition to the items on the guides, other small moving parts, such as hinges and latches, must be lubricated at frequent intervals.

b. References. Materiel must be lubricated in accordance with the latest instructions contained in technical manuals and/or ordnance field service bulletins. Reference is made to the General Instruction section (OFSB 6-4) for additional lubrication information, and to the Product Guide section (OFSB 6-2) for latest approved lubricants.

c. Supplies. In the field, it may not be possible to supply a complete assortment of lubricants called for by the lubrication guide to meet the recommendations. It will be necessary to make the best use of those available, subject to inspection by the officer concerned, in consultation with responsible ordnance personnel.

d. Lubrication Notes. The following notes apply to the lubrication guides (figs. 50 and 51). All note references in the guides themselves are to the following subparagraphs having the corresponding numbers:

(1) FITTINGS AND OILERS. Clean before applying lubricant. Where bearing can be seen, lubricate until new lubricant is forced from bearing.
CAUTION: Lubricate after washing materiel.

(2) INTERVALS. Intervals indicated are for normal service. For extreme conditions of speed, heat, water, sand, mud, snow, rough roads, and dust, lubricate more frequently.

(3) RECOIL FLUID. For instructions on quantity and application or recoil fluid, refer to War Department Recoil Fluid Guide No. 19 and OFSB 6-6.

TM 9-331
155-MM HOWITZER M1 AND 155-MM HOWITZER CARRIAGE M1

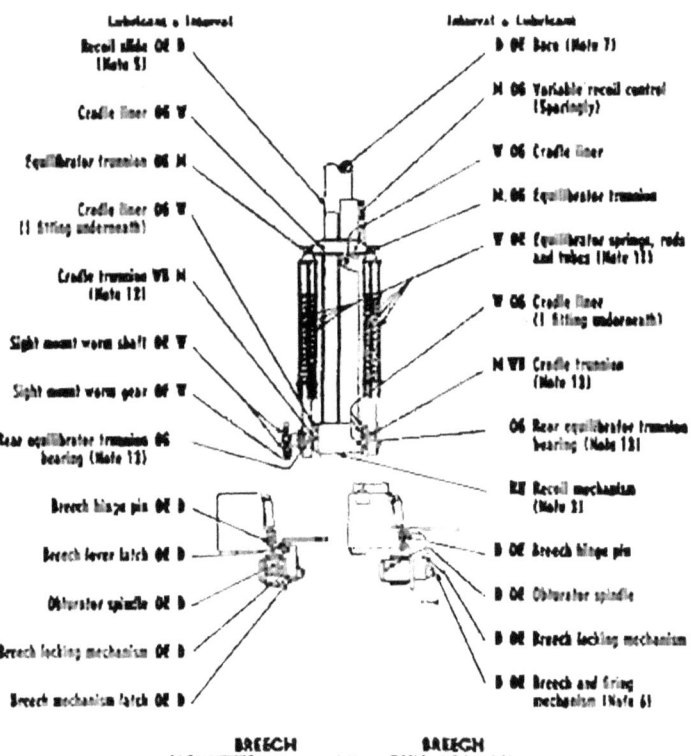

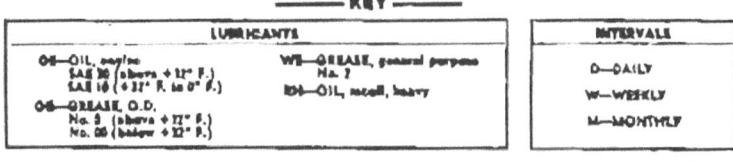

Figure 50—Lubrication Guide for 155-mm Howitzer M1 and Recoil Mechanism

TM 9-331
45

LUBRICATION

(4) TRAVERSING AND ELEVATING WORM GEAR CASES. Monthly, check level; if necessary, add lubricant to correct level. Every 6 months, drain, flush, and refill.

(5) RECOIL SLIDE. Daily, and before firing, clean and oil exposed metal surfaces. Keep exposed surfaces covered with thin film of OIL, lubricating, engine (seasonal grade).

(6) BREECH AND FIRING MECHANISM. Daily, and before and after firing, clean and oil all moving parts and exposed metal surfaces with, OIL, lubricating, engine (seasonal grade).

CAUTION: To insure easy breech operation and to avoid misfiring in cold weather, clean with SOLVENT, dry-cleaning; dry and lubricate with OIL, lubricating, preservative, light. To clean firing mechanism, remove mechanism and operate pin in SOLVENT, dry-cleaning.

(7) HOWITZER BORE. Daily, and after firing, clean and coat with OIL, lubricating, engine (seasonal grade).

(8) WHEEL BEARINGS. Remove wheel; clean and repack bearings. To clean and pack wheel bearings properly, they must be removed from the hub. Follow this procedure:

(a) Remove the bearings from the hub and wash in SOLVENT, dry-cleaning, until all the old lubricant is removed from both inside and outside of the cage.

(b) Lay bearings aside to dry and wash inside of hub and spindle with SOLVENT, dry-cleaning.

(c) When bearings are thoroughly dry, coat lightly with OIL, lubricating, engine (seasonal grade) and pack races with GREASE, general purpose, No. 2, and reassemble in hub. To pack a bearing satisfactorily, it is necessary to knead lubricant by hand into space between the cage and inner race. Do not apply any lubricant to inside of hub or on spindle. The lubricant packed in the bearing races is sufficient to provide lubrication until the next service period. An excess may result in leakage of the lubricant into the brake drum.

(d) Mount wheel on spindle and tighten nut on end of spindle until there is a slight drag when wheel is rotated.

(e) Back off nut until wheel turns freely without side play. Lock adjusting nut in position.

(f) Install hub plate. Lubricate bearings only.

(9) TRAVERSING AND ELEVATING RACKS AND PINIONS. Daily clean and apply OIL, lubricating, engine (seasonal grade).

(a) The teeth of the traversing and elevating arcs require little lubrication, but as a protection against rust, they must be covered with a thin coat of oil. Dust and grit will adhere to this oily film. Consequently, the teeth must be thoroughly clean and fresh oil applied before traversing or

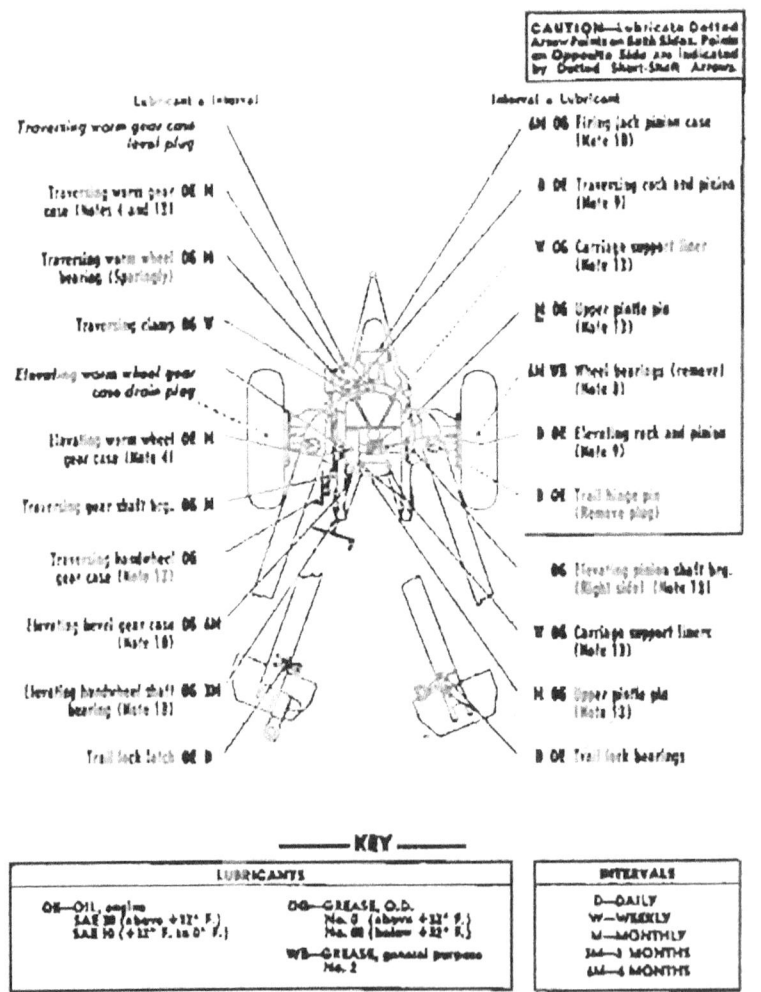

Figure 51—Lubrication Guide for 155-mm Howitzer Carriage M1

LUBRICATION

elevating the gun; otherwise the grit will cause rapid wear of the arc and pinion gear. If considerable dust is present when gun is operated, the oil should be removed from the gear teeth. The teeth should remain dry until action is over. If the surfaces are dry, there is less wear than when they are coated with lubricant contaminated with grit.

(10) FIRING JACK PINION AND ELEVATING BEVEL GEAR CASES. Every 6 months, remove cover plate, clean rack and gear teeth with SOLVENT, dry-cleaning, and repack with GREASE, O.D. (seasonal grade).

(11) EQUILIBRATOR SPRINGS, RODS, AND TUBES. Weekly, clean and apply OIL, lubricating, engine (seasonal grade).

(12) OILCAN POINTS. Daily, lubricate gas check pad and split rings with OIL, lubricating, engine (seasonal grade).

(13) POINTS TO BE SERVICED AND OR LUBRICATED BY ORDNANCE MAINTENANCE PERSONNEL AT TIME OF ORDNANCE INSPECTION.—Cradle trunnion bearings, traversing handwheel gear case, elevating pinion shaft bearing (right side), traversing worm gear case, traversing mechanism spur gear shaft bearing, carriage support liners, lower pintle bearings, rear equilibrator trunnion bearings, elevating handwheel shaft bearing.

46. REPORTS AND RECORDS.

a. Reports. If lubrication instructions are closely followed, proper lubricants used, and satisfactory results are not obtained, a report will be made to the ordnance officer responsible for the maintenance of the materiel.

b. A complete record of lubrication servicing will be kept in the Artillery Gun Book (O.O. Form 5825) for the materiel.

TM 9-331
47

155-MM HOWITZER M1 AND 155-MM HOWITZER CARRIAGE M1

Section VI

CARE AND PRESERVATION

	Paragraph
General	47
Organization spare parts and accessories	48
Cleaning after firing	49
Howitzer	50
Recoil mechanism	51
Carriage	52
Cleaners and abrasives	53
Preservatives	54
Miscellaneous materials and tools	55
Washing	56
Painting	57

47. GENERAL.

a. It is of vital importance that all parts of the materiel be kept in proper condition in order that the weapon be ready for immediate service. Lubricating, cleaning, and preserving materials provided with the howitzer and carriage will enable the personnel to keep the parts in proper working condition. This section of the manual prescribes the uses of these materiels.

b. Moving parts of the various mechanisms should be lubricated in the prescribed manner, and periodical examinations should be made to insure that the lubricant is reaching the parts for which it is intended, and to insure that the weapon as a whole is receiving the attention necessary for satisfactory operation.

c. Dirt and grit, accumulated in traveling or from the blast of the piece in firing, settle on bearing surfaces, and in combination with the lubricant itself, form a cutting compound. Powder fouling attracts moisture and hastens the formation of rust. It is essential that all parts be cleaned at frequent intervals, depending upon use and service.

d. If rust should accumulate, its removal from bearing surfaces requires special care in order that clearances shall not be unduly increased. CLOTH, crocus, should be used for this purpose. The use of coarse abrasives is strictly forbidden.

e. In disassembly, assembly, or inspection, extreme care must be exercised to prevent dust, dirt, or other foreign matter from entering the mechanisms.

CARE AND PRESERVATION

f. When materiel is not in use, the proper covers must be used.

g. When the weapon is to be unused for a considerable time, the bore, breech mechanism, and bright unpainted surfaces should be cleaned with SOLVENT, dry-cleaning, and the surfaces coated with COMPOUND, rust-preventive.

h. Should an enemy shell burst near the weapon, it must be determined that the weapon has not been damaged to a dangerous degree before the next round is fired. Damage of a serious nature should be reported to the ordnance officer.

48. ORGANIZATION SPARE PARTS AND ACCESSORIES.

a. All organization spare parts, tools, and accessories should be kept in an orderly manner so that they can be quickly located when required. They should be protected from loss or damage by being kept in their proper chests. Those items susceptible to rust and corrosion must be cleaned thoroughly at regular intervals and coated with a film of oil. Parts supplied in protective containers should be kept in the containers until required.

b. The sets of organization spare parts and accessories for the gun and carriage should be maintained as completely as possible at all times. The sets should be checked with the lists in the standard nomenclature lists (chap. 9), and all used parts, and all missing parts, tools, and accessories should be replaced immediately.

49. CLEANING AFTER FIRING.

a. Bore. This procedure is to be followed immediately after firing and while the howitzer is still warm. It is to be repeated at the end of 24 hours and again at the end of 48 hours. The purpose of the second and third washing, drying, and oiling operations is to remove the effects of "sweating," a chemcial reaction of the burned powder which cannot be removed with the initial procedure.

(1) CLEAN BORE. Assemble the staff and attach the cleaning and unloading Rammer M7. Wrap the head with pieces of burlap soaked in a solution made by dissolving ½ pound of SODA ASH, or 1 pound of sal soda, in a gallon of water. With 4 or 5 men on the staff, work the head through the bore, using a pushing and pulling action. A man posted at the muzzle of the howitzer can indicate by movements of his hands the progress of the head in the bore and prevent its slipping from the muzzle of the howitzer.

(2) CLEAN CHAMBER. Use the cleaning and unloading Rammer M7 on a single section of the staff. Wrap the head with burlap soaked in the solution. Use another section of the staff or an axe handle as a

TM 9-331
49-50

155-MM HOWITZER M1 AND 155-MM HOWITZER CARRIAGE M1

lever to hold the cleaning element against the sides of the chamber. Pressure must be applied to clean thoroughly.

(3) DRY BORE AND CHAMBER. Follow the procedure given for cleaning, using the same accessories and methods, but substituting dry burlap for burlap soaked in the solution.

(4) OIL BORE. Assemble the Bore Brush M22 to the staff. Place the brush over a bucket to catch the oil which drips from the brush, and apply oil to the brush, using a paint brush for this purpose. Use OIL, engine, lubricating (seasonal grade). Apply a light coating of oil to the bore by working the brush through the bore.

(5) OIL CHAMBER. The chamber and threads of the breech ring may be thoroughly coated with oil by using a paint brush nailed at right angles to the end of a 5-foot, round-cornered piece of wood. The long handle will permit reaching all parts of the breech recess and doing a good job with a minimum of waste.

b. Breech Mechanism. After firing, the breech mechanism should be disassembled (par. 66), cleaned, oiled, and assembled. Any malfunction or deformation of parts should be reported to ordnance maintenance personnel for correction. Particular care should be given the cleaning of the vent through the obturator spindle and the primer seat. During firing, residue works its way back through the venthole in the spindle to the primer seat. If this is not removed, the primer will seat imperfectly. If the vent is not free and clear, the flash from the primer cannot properly ignite the charge.

c. Firing Mechanism M1. After firing, the firing mechanism should be disassembled (par. 64), cleaned, oiled, and assembled. Special care must be exercised to keep this mechanism well oiled and free from rust, powder stains, and dirt.

50. HOWITZER.

a. General.

(1) It is known that the wear on cannon does not depend entirely upon the number of rounds fired, but very much upon the care taken to allow cooling between rounds and upon a thorough cleaning and oiling schedule followed through in a consistent manner. It is essential that every projectile be cleaned thoroughly before it is inserted in the weapon.

(2) The muzzle cover must be kept in place when the howitzer is not in action.

(3) The surfaces of the leveling plates should be protected from injury. Any necessary repairs will be made by ordnance maintenance personnel.

CARE AND PRESERVATION

(4) It is important that any cutting or abrasion of the threads or bearing surfaces of the breechblock or breech ring be reported to ordnance maintenance personnel for correction.

(5) As the accuracy life of cannon is decreased by a fast rate of fire and the attendant heat, the piece should be washed, oiled, and allowed to cool as often as is practicable.

b. Regular Cleaning Procedure. This cleaning procedure is to be followed at intervals specified by the officer in charge. The interval will be dependent upon atmospheric, traveling, or other conditions.

(1) CLEAN BORE AND CHAMBER. Swab the bore and chamber in the manner described in paragraph 49 a (1) and (2), using SOLVENT, dry-cleaning, in place of the SODA ASH or sal soda solutions.

(2) DRY BORE AND CHAMBER. Dry the bore and chamber in the manner described in paragraph 49 a (3).

(3) OIL BORE AND CHAMBER. Oil the bore and chamber in the manner described in paragraph 49 a (4) and (5).

c. Inactivity. If the weapon is to be unused for a considerable length of time, the bore should be given a coating of heavy rust preventive.

d. Breech Mechanism.

(1) At frequent intervals, when the gun is not being fired, the breech and firing mechanisms will be disassembled, cleaned, and oiled. These intervals will be specified by the officer in charge, and their frequency will be dependent upon atmospheric and other conditions.

(2) When the howitzer is not in use, the breech cover will be in place to prevent dust and grit from getting into the recesses of the mechanism and impeding its easy operation.

(3) If the breechblock does not rotate smoothly, or if the mechanism requires greater effort than usual to operate, this should be considered sufficient warning to warrant disassembly of the breech mechanism (par. 66) to determine the cause. If corrective measures are beyond the scope of the using force, the matter should be brought to the attention of ordnance maintenance personnel.

e. Gas Check Seat. Extreme care must be taken to prevent injury to the gas check seat as this affects the seating of the split rings and is likely to cause leakage of gas and burning of the gas check pad. A leak may result in serious erosion. Because the gas check is susceptible to moisture, rusting is likely to occur. The seat and rings, therefore, should be well protected by oil at all times, and by COMPOUND, rust-preventive, when the howitzer is not in use.

TM 9-331
50-51

155-MM HOWITZER M1 AND 155-MM HOWITZER CARRIAGE M1

NOTE: The gas check seat is that portion of the rear of the chamber on which the split rings and the gas check pad bear when the breech is in sealed position and the howitzer is fired.

f. **Gas Check Pad.** The gas check pad must be kept in a condition that will permit its being compressed without injury. It should be oiled frequently with OIL, engine, SAE 10, lubricating, and the oil should be rubbed in well. A pad of the proper resiliency will yield slightly under heavy pressure from the thumbs.

g. **Firing Mechanism M1.** Except when the howitzer is loaded and is being fired, the Firing Mechanism M1 will be removed from the howitzer. When not in use, it will be kept in the protected place reserved for its stowage. The percussion hammer will be pinned in the upright position by raising the hammer and turning the hammer latch knob so that the latch pin enters its recess in the hammer. The lanyard will be detached and the percussion hammer will not be unnecessarily brought in contact with the firing pin or the firing mechanism housing. Bringing the hammer into contact with the firing mechanism housing will result in damage to the threads of the housing.

51. RECOIL MECHANISM.

a. General.

(1) The replenisher and the recuperator must contain the proper oil reserves at all times or damage to the materiel will result when the howitzer is fired. The oil reserve in the replenisher supplies any needed recoil oil to the recoil cylinder during the recoil of the howitzer. It also compensates for expansion and contraction of the recoil oil due to heat and cold. The oil reserve in the recuperator separates the floating piston from the regulator valve.

(2) The compressed nitrogen in the recuperator will put pressure on the oil in the counterrecoil system only so long as there is oil between the regulator and the floating piston. In the event that these pieces come in contact, further movement of the floating piston is prevented and the pressure on the oil drops to zero.

(3) The recoil mechanism should be examined regularly for leakage of oil. There is no cause for alarm should the oil drip rapidly, or even run in a stream, from the rear of the replenisher when the howitzer is elevated, provided that the weapon has been at zero elevation for some time. This condition may exist in a normal replenisher. A leak at any packing that does not exceed 3 drops per minute is not considered serious; if the loss of oil is greater, report the matter to the ordnance maintenance personnel.

TM 9-331
51

CARE AND PRESERVATION

(4) Keep the air holes in the replenisher open. The two air holes are small continuations of the tent wrench holes in the replenisher piston guide and provide for the circulation of air at the rear of the replenisher piston. Use a wire for cleaning (wood may splinter and stop up the hole).

(5) Use a screwdriver to remove and replace the replenisher filling and drain plug. Use a ½-inch socket wrench (hexagonal) to remove the recuperator filling and drain plug to avoid injury of the head. Be careful that the gasket beneath the plug is not lost. Examine the threads of the plug before replacing it. If the threads are in good condition, screw the plug up on the gasket, but do not force it excessively.

b. Recoil Oil.

(1) OIL, recoil, heavy, is used in the recoil mechanism. Care must be taken not to use recoil oil other than that prescribed. Water and other foreign matter must not be introduced into recoil mechanisms that use recoil oil.

(2) Recoil oil must not be put into any container not marked with the name of the oil, nor left in open containers, nor be subjected to excessive heat, nor mixed with any other type oil. The transfer of recoil oil to a container not marked with the name of the oil may result in the wrong oil getting into the recoil mechanism, or in the use of recoil oil for lubricating purposes.

(3) When putting recoil oil in the system, it should be filtered through a piece of clean cloth as well as through the wire strainer of the filling funnel. Every precaution must be taken to prevent the introduction of water or grit into the mechanism, either in the oil or through failure to clean thoroughly the connections and servicing equipment.

(4) Exposure of recoil oil in an open can may result in the accumulation of moisture. Condensation in a container partly filled with oil, or the pouring of oil from one container to another which has moisture on its inner walls, results in moisture being carried along with the oil into the recoil mechanism.

(5) If there is a possibility that recoil oil may contain water, it should be tested by one of these methods. Fill a clean glass container of pint capacity with the recoil oil. Permit the oil to settle. The water, being heavier than the oil, will sink to the bottom, if present. With the container slightly tilted, drops or bubbles will form in the lower portion. Invert the container and hold it to the light. Drops or bubbles of water, if present, may be seen slowly sinking in the oil. If the oil has a cloudy appearance, the cloudiness may be ascribed to particles of water.

(6) Another test for water is to heat the oil to 212 F (boiling point of water) in a shallow pan. Water in the oil will appear on the surface as minute bubbles. This test will disclose water not determinable by the

83

TM 9-331
51

155-MM HOWITZER M1 AND 155-MM HOWITZER CARRIAGE M1

Figure 52 — Establishing Replenisher Oil Reserve

settling test. Should either test show water, the oil on hand should be turned in.

NOTE: Purging means removing all air from the filling device before forcing oil into the recoil mechanism. This is accomplished by only partially screwing home the connection to the mechanism, operating the filling device, and letting the air in the filling device escape until no more bubbles appear at the connection, then tightening the connection fully.

c. To Test Operation of Replenisher Piston.

(1) Remove the plug from the rear of the replenisher and insert a scale through the opening of the replenisher piston guide and against the rear end of the piston extension. Then release oil from the replenisher by screwing the filling and drain valve oil release into the filling and drain valve located on the front of the replenisher housing (fig. 52 b). If movement of the replenisher piston takes place, the replenisher piston is functioning.

(2) If the replenisher piston does not move, insert a block of hardwood through the opening in the piston guide against the piston end and tap with a hammer. When the replenisher has not been exercised, the

CARE AND PRESERVATION

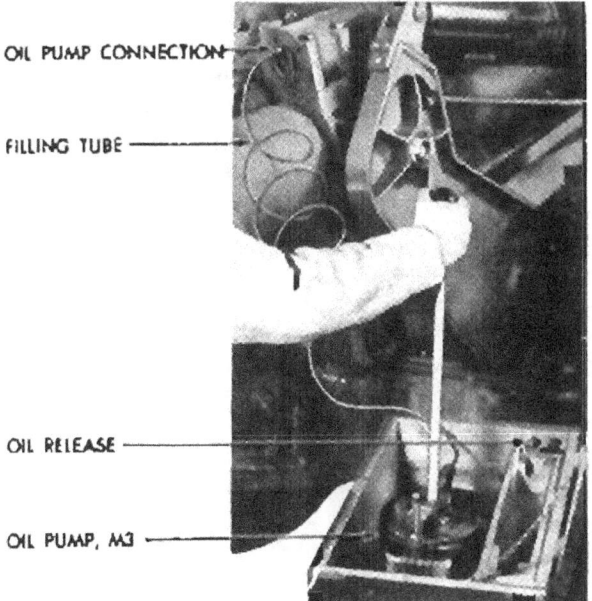

Figure 53—Establishing Recuperator Oil Reserve

piston extension may become rusted in the guide. Any rust on the extension should be removed.

  d. **To Establish Replenisher Oil Reserve with Oil Pump M3.**

  (1) The position of the replenisher piston extension in the piston guide governs the filling of the replenisher (par. 29 b). Before filling the replenisher, test the operation of the replenisher piston in the manner described in the preceding paragraph.

  (2) Unscrew the filling and drain plug from the filling and drain valve on the front left side of the replenisher, and screw the connection of the filling tube loosely into the filling hole. Purge the tube and tighten the connection without the use of a wrench except for the final tightening. Extreme care must be taken to prevent injury to the threads of the filling hole as any damage may put the entire weapon out of commission until repaired.

  (3) Work the pump (fig. 52 a) until the rear end of the replenisher piston extension is 5½ inches (140-mm) from the rear face of the replenisher. Remove the filling tube connection and replace the filling and drain plugs.

155-MM HOWITZER M1, AND 155-MM HOWITZER CARRIAGE M1

e. To Exercise Replenisher Piston.

(1) The replenisher may be exercised by pumping oil into the replenisher until the rear end of the piston extension projects to the rear of the replenisher. Any visible rust should be polished off the extension. Enough oil should then be withdrawn by means of the filling and drain valve oil release to bring the piston back to normal position.

(2) Howitzers not being fired should be exercised in the manner prescribed at least once a month.

f. To Establish Recuperator Oil Reserve with Oil Pump M3.

(1) The position of the oil index (par. 29 c) indicates the oil reserve in the recuperator. When the counterrecoil of the howitzer, or the position of the oil index, indicates that there is too small a quantity of oil in the recuperator, it will be necessary to drain off the reserve oil before refilling. This is accomplished by removing the filling and drain plug and inserting the filling and drain valve oil release in the recuperator rear cylinder head.

(2) The reserve oil will spurt out in a stream and suddenly drop to a trickle. The amount of reserve oil which will escape will be approximately 1 quart. At this point, the flow of oil should be stopped by unscrewing the oil release. It will be noted that the oil index has moved out of sight before all of the reserve oil has been released. If the oil index has not moved, tap it gently with a small piece of wood as it may be frozen.

(3) To replenish the reserve, clean the connection of the coil of the Oil Pump M3, screw the connection loosely in the filling and drain valve, purge the pump, tubing, and connection, and tighten the connection in the filling and drain valve.

(4) Start working the oil pump while closely watching the oil index. As soon as the oil index starts to move outwardly, begin counting the strokes of the pump handle. Count the number of strokes required to bring the oil index to its farthest outward position. Multiply the number of strokes required to accomplish this result by *three*. Add this number of strokes to the oil reserve. This will constitute a full reserve. Detach the connection and replace the plug.

52. CARRIAGE.

a. General.

(1) The care and preservation of the carriage in service requires proper cleaning, strict observance of the lubrication program, tightening of loosened parts, and repair or replacement of broken parts. When traveling, it calls for proper attachment of the traveling lock, secure locking of the trail lock, firing jack, and accessories in their positions on the carriage, the full protective use of the covers, the proper adjustment of brakes, and the correct inflation of tires.

CARE AND PRESERVATION

(2) All bearing surfaces, screw threads, and exterior parts must be clean and as free from dirt as possible. Special attention should be given exposed teeth and bearing surfaces. Cotter pins must be properly spread. Bolts, nuts, and screws must be tight and locked to prevent their coming loose. The best available tools should be used.

(3) The carriage should receive a general inspection periodically.

(4) When the carriage is stored, or is not to be used for a considerable length of time, all bright and unpainted surfaces must be protected with COMPOUND, rust-preventive. They should first be cleaned with SOLVENT, dry-cleaning.

(5) Vigilance must be exercised to detect any cutting or abrasion of the teeth of the pinions and arcs of the elevating and traversing mechanisms, or on the plunger and pinion of the firing jack. Any deformation of this nature should be reported to the ordnance maintenance personnel for correction. Rust must not be permitted to accumulate on these parts.

b. Pneumatic Tires and Tubes.

(1) To obtain maximum mileage, the air pressure in the tires should be checked prior to operations and thereafter left alone unless there is a loss of air pressure. Bleeding of air from the tires results in an increase of the flexing of the tire side walls which increases the dangers of the failure.

(2) Remove all foreign substances from the rubber, being especially careful to keep tires as free from oil and grease as possible. Oil and grease have a deteriorating effect upon rubber.

c. Care of Electric Brakes. Proper attention must be given the electrical brake system to insure proper functioning. It is essential that the using force has knowledge which will permit the correction of malfunctions which may occur during traveling. This knowledge will also permit the best possible service from this equipment. The following information and that contained in paragraph 43 a through e will prove helpful:

(1) WIRING AND CONTROLLER. Check the wiring and the controller before examining the brakes. When the brakes are new, several applications must be made before maximum efficiency is obtained.

(2) FACING OF MAGNET. The facing of the magnet may become glazed. This is not characteristic of the facing, but is due to some foreign substance embedding itself into the material, resulting in a polished surface. If the facing cannot be roughened with coarse emery cloth, notify ordnance maintenance personnel.

(3) CURRENT. Check current at the brakes, using the ammeter furnished as an accessory. Disconnect one brake wire only. Connect one side of the ammeter to the brake; the other side to the terminal of the

TM 9-331
52

155-MM HOWITZER M1 AND 155-MM HOWITZER CARRIAGE M1

Figure 54 — Testing Current at Brakes

live wire that was removed from the brake. Leave the other brake in the circuit (fig. 54). Take a reading; it should not be less than 2.6 amperes. If this amount of current is not available, the brakes will not operate properly.

(4) Check for poor connections and partly broken or worn wires. Check current consumption of the brake on the opposite side, with the first brake connected in the circuit. The reading of both magnets should not vary more than 0.1 of an ampere. In case there is a greater variation, check all connections for poor contact or for a broken wire at the magnet. Test the battery of the prime mover to see if it is sufficiently charged to turn over the starting motor. When the test is completed, remove the ammeter from the circuit and connect the live wire to the brake terminal.

(5) STOP LIGHTS. Stop lights must not be connected into the brake circuit. This changes the amount of the current which passes through the controller, resulting in weak or grabbing brakes.

(6) BEARINGS AND WHEELS. Worn bearings or loose wheels will cause erratic action of the brakes and will be evidenced by the wide track the pole faces of the magnets make on the armatures. The roller bearings must be adjusted, or broken or badly worn bearings must be replaced.

CARE AND PRESERVATION

(7) BRAKE LINING. Notify ordnance maintenance personnel when brake lining is worn out or has become greasy.

(8) BRAKE DRUMS OUT OF ROUND. Notify ordnance maintenance personnel.

(9) BRAKE BAND DISTORTED. Notify ordnance maintenance personnel.

(10) BROKEN OR WEAK BRAKE BAND SPRING. Remove the wheel and replace the spring.

(11) CONTROLLER BURNED OUT. Replace with a new controller.

(12) BROKEN MAGNET SPRING. Remove the wheel and replace the spring.

(13) BUSHING IN MAGNET WORN OUT. Notify ordnance maintenance personnel.

(14) BENT CONTACTOR BLADES IN CONTROLLER. Straighten blade with flat-nose pliers to correct spacing.

(15) BROKEN SPRING IN HAND CONTROL. Replace with new spring.

(16) WARPED BACKING PLATE. Notify ordnance maintenance personnel.

(17) INSUFFICIENT SPACING BETWEEN ARMATURE AND MAGNET. Check with the armature gage. Notify ordnance maintenance personnel.

53. CLEANERS AND ABRASIVES.

a. The following cleaners and abrasives are prescribed for use with this materiel. See TM 9-850 for full information.

BURLAP, jute, 8-oz, (40 in. wide)
CLOTH, crocus
CLOTH, emery
CLOTH, wiping, cotton, mixed, sterilized (for machinery)
COMPOUND, cleaning, trisodium phosphate
COMPOUND, rust-preventive, heavy
COMPOUND, rust-preventive, light
LIME, hydrated
PAPER, flint
PAPER, lens, tissue
POLISH, metal, paste
REMOVER, paint and varnish
SOAP, saddle
SODA ASH
SODA, caustic (lye), for cleaning purposes
SOLVENT, dry-cleaning
SPONGES
WASTE, cotton (two grades, colored and white)

54. PRESERVATIVES.

a. See TM 9-850 for information on rust, corrosion, inspection for corrosion, rust prevention, preparation of metal surfaces for slushing, method of slushing, inspection of grease films, and storage conditions.

155-MM HOWITZER M1 AND 155-MM HOWITZER CARRIAGE M1

h. Naphthalene, Flake.

(1) A flaked form of moth ball, used as a moth repellent to preserve the linings of helmets, felt wads, felt packings of instrument chests, carpets, howitzer sponges, and paint and varnish brushes. It is sprinkled thickly on the articles which should, if possible, be then wrapped in paper covers and tightly boxed. The materiel should be thoroughly brushed and aired before packing and should be periodically inspected. If there are any signs of destruction by the moth larvae, the articles must be unpacked, cleaned, and recharged with naphthalene.

(2) Naphthalene should be used in airtight receptacles in order to obtain a concentrated naphthalene vapor.

55. MISCELLANEOUS MATERIALS AND TOOLS.

a. For the purpose for which the following materials and tools are used, see TM 9-850.

BRUSH:
  Artist, camel's-hair, rd., No. 1
  Flowing, skunk's-hair, No. 3 (2-in.)
  Sash-tool, oval, No. 1 ($^{23}/_{32}$- x 1¼-in.)
  Sash-tool, oval, No. 3 (1$^{3}/_{32}$- x 2½-in.)
  Scratch, wire, painter's (7¼- x 2¾-in.) curved back (14- x ⅜-in.) handled
  Varnish, oval (1⅞-in.)
CHALK, white, railroad (1- x 4-in.)
NEEDLE, sacking, steel (4½-in.)
PALM, sailmaker's
TWINE, jute

b. Care of Brushes.

(1) Brushes are subject to attack by moths. Brushes in storage should be protected by naphthalene.

(2) Camel's-hair brushes, after being thoroughly cleaned with turpentine, should be laid flat on a horizontal surface (not in water). For temporary storage, other paint brushes should be cleaned after using and kept with bristles submerged in fresh water, being careful that the bristles do not touch the bottom of the container.

56. WASHING.

a. Serious damage to ordnance materiel, in many cases requiring repair and replacement of component parts of sighting equipment, fire-control instruments, and weapons and carriage, has frequently resulted from the use of water, steam, or air from a high-pressure hose for cleaning purposes. For this reason, operating personnel is cautioned to prevent

CARE AND PRESERVATION

water, dirt, or grit from being forced into the cradle, yoke, or cover, trunnion bearing or trail hinge pin housings, bearing surfaces, or gear cases when using water, steam, or air under pressure for cleaning.

b. Under no circumstances will a hose, either normal-pressure or high-pressure, be used in cleaning any sighting equipment of any fire-control instruments. Before washing, take off removable sighting equipment from the materiel to be cleaned. In cases where it is not removable, take care to cover the parts properly.

57. PAINTING.

a. General.

(1) Ordnance materiel is painted before being issued to the using arms and one maintenance coat per year will ordinarily be ample for protection. With few exceptions this materiel will be painted with ENAMEL, synthetic, olive-drab, lusterless. The enamel may be applied over old coats of long-oil enamel and oil paint previously issued by the Ordnance Department if the old coat is in satisfactory condition for repainting.

(2) Paints and enamels are usually issued ready for use and are applied by brush or spray. They may be brushed on satisfactorily when used unthinned in the original package consistency, or when thinned no more than 5 percent by volume with THINNER. The enamel will spray satisfactorily when thinned with 15 percent by volume of this thinner. (Linseed oil must not be used as a thinner since it will impart a luster not desired in this enamel.) If sprayed, it dries hard enough for repainting within ½ hour and dries hard in 16 hours.

(3) Certain exceptions to the regulations concerning painting exist. Fire-control instruments, for instance, which require a crystalline finish, will not be painted with ENAMEL, synthetic, olive-drab, lusterless.

(4) Complete information on painting is contained in TM 9-850.

b. Preparation for Painting.

(1) If the base coat on the materiel is in poor condition and it is desirable to strip the old paint from the surface rather than to use sanding and touch-up methods, it will be necessary to apply a primer coat.

(2) PRIMER, synthetic, refinishing, should be used on wood as a base coat for synthetic enamel. It may be applied either by brushing or spraying. It will brush satisfactorily as received, or after the addition of not more than 5 percent by volume of THINNER. It will be dry enough to touch in 30 minutes, and hard in 5 to 7 hours. For spraying, it may be thinned with not more than 15 percent by volume of THINNER. Lacquers must not be applied to the PRIMER, synthetic, refinishing, within less than 48 hours.

TM 9-331
57

155-MM HOWITZER M1 AND 155-MM HOWITZER CARRIAGE M1

(3) PRIMER, synthetic, rust-inhibiting, for bare metal, should be used on metal as a base coat. Its use and application is similar to that outlined in subparagraph (2), above.

(4) The success of a job of painting depends partly on the selection of a suitable paint, but also largely upon the care used in preparing the surface prior to painting. All parts to be painted should be free from rust, dirt, grease, kerosene, and alkali, and must be dry.

c. *Painting Metal Surfaces.*

(1) Metal parts may be washed in a liquid solution consisting of ½ pound of SODA ASH in 8 quarts of warm water, then rinsed in clear water and wiped thoroughly dry. Wood parts may be treated in the same manner but the alkaline solution must not be left on for more than a few minutes and the surfaces should be wiped dry as soon as they are washed clean.

(2) When artillery or automotive equipment is in fair condition and only marred in spots, the bad places should be touched with ENAMEL, synthetic, olive-drab, lusterless, and permitted to dry. The whole surface should then be sandpapered with PAPER, flint, No. 1, and a finish coat of ENAMEL, synthetic, olive-drab, lusterless, applied and allowed to dry thoroughly before the materiel is used.

(3) If the equipment is in bad condition, all parts should be thoroughly sanded with PAPER, flint, No. 2, given a coat of PRIMER, synthetic, refinishing, and permitted to dry for at least 16 hours. Sandpaper with PAPER, flint, No. 00, wipe free from dust and dirt, and apply a final coat of ENAMEL, synthetic, olive-drab, lusterless; allow the materiel to dry thoroughly before it is used.

d. *Paint as a Camouflage.* Camouflage is now the major consideration in painting ordnance materiel, with rust prevention secondary. The camouflage plan at present employed utilizes color and gloss.

(1) COLOR. Materiel is painted with ENAMEL, synthetic, olive-drab, lusterless, which was chosen to blend in reasonably well with the average landscape.

(2) GLOSS. The new lusterless enamel makes the materiel difficult to see from the air or from great distances over land. Materiel painted with ordinary glossy paint can be detected more easily and at greater distances.

(3) PRESERVING CAMOUFLAGE.

(*a*) Since continued friction or rubbing will smooth the surface and produce a gloss, it must be avoided. The materiel should not be washed more than once a week. Care should be taken to see that the washing is done entirely with a sponge or a soft rag. The surface should never be rubbed or wiped, except while wet, or a gloss will be developed.

(*b*) It is not desirable that materiel painted with lusterless enamel be kept as clean as when glossy paint was used. A small amount of dust

CARE AND PRESERVATION

increases the camouflage value. Grease spots should be removed with SOLVENT, dry-cleaning. Whatever portion of the spot cannot be so removed should be allowed to remain.

(c) Continued friction of wax-treated tarpaulins on the sides of the materiel will also produce a gloss. It should be removed with SOLVENT, dry-cleaning.

(d) Tests indicate that repainting will be necessary once yearly in the case of the olive-drab and twice yearly in the case of the blue-drab enamel.

e. Removing Paint.

(1) After repeated paintings, the paint may become so thick as to scale off in places and present an unsightly appearance. If such is the case, remove the old paint by use of a lime-and-lye solution or REMOVER, paint and varnish. It is important that every trace of lye or other paint remover be completely rinsed off and that the equipment be perfectly dry before repainting is attempted. It is preferable that the use of lye solutions be limited to iron or steel parts.

(2) If used on wood, the lye solution must not be allowed to remain on the surface for more than a minute before being thoroughly rinsed off and the surface wiped dry with rags. Crevices or cracks in wood should be filled with putty and the wood sandpapered before finishing. The surfaces thus prepared should be painted according to the directions given previously.

f. Painting Lubricating Devices. Oil cups, grease gun fittings, oilholes, and similar lubricating devices, as well as a circle about ¾ inch in diameter at each point of lubrication will be painted with ENAMEL, synthetic, gloss-red, in order that they may be readily located.

TM 9-331
58-59

155-MM HOWITZER M1 AND 155-MM HOWITZER CARRIAGE M1

Section VII

INSPECTION AND ADJUSTMENT

	Paragraph
Purpose	58
Visual inspection upon receipt	59
Serial numbers	60
Inspection of howitzer	61
Inspection of carriage	62

58. PURPOSE.

a. Inspection has as its purpose the detection of conditions which might cause improper performance. Such conditions may be caused by:

(1) Mechanical deficiencies resulting from ordinary wear, breakage, or exposure to the elements or enemy fire.

(2) Faulty or careless operation.

(3) Improper care, such as inadequate lubrication, inadequate protection, or insufficient preservative measures.

b. Inspection should always be accompanied by corrective measures to remedy any deficiencies found. When properly carried out, inspection and necessary corrective maintenance will insure the maximum reliability and performance of the materiel. The inspection outlined in this section of this manual should be made at regular intervals of not less than 30 days during both active and inactive seasons.

c. Before inspecting particular points, the howitzer and carriage should be inspected in general for evidences of faulty operation, care, or maintenance. Any unusual conditions which might result in improper operation or damage to the materiel will be immediately remedied. Untidy appearance and evidences of rust or deterioration will be corrected. Missing or broken apparatus will be replaced.

59. VISUAL INSPECTION UPON RECEIPT.

a. Upon receipt of this materiel, it is the responsibility of the officer in charge to ascertain whether it is complete and in sound operating condition. A record should be made of any missing parts and of any malfunctions, and any such conditions should be corrected as quickly as possible.

b. Attention should be given to small and minor parts as these are the more likely to become lost and may seriously affect the proper functioning of the materiel.

c. This visual inspection upon receipt should be followed as quickly as possible by a complete inspection which will disclose the fuctioning of the materiel (pars. 61 and 62).

TM 9-331
59-61

INSPECTION AND ADJUSTMENT

Figure 55—Recoil Mechanism Serial Number

60. SERIAL NUMBERS.

a. Three serial numbers are required for records concerning the components of this materiel. They are the howitzer serial number, the recoil mechanism serial number, and the carriage serial number.

b. *Howitzer Serial Number.* Stamped on top of the breech ring (fig. 6).

c. *Recoil Mechanism Serial Number.* On brass plate on right side of recoil mechanism cover near yoke (fig. 55).

d. *Carriage Serial Number.* On brass plate on right arm of top carriage (fig. 56).

61. INSPECTION OF HOWITZER.

a. The following instructions will be scrupulously observed by all concerned:

Figure 56—Carriage Serial Number

155-MM HOWITZER M1 AND 155-MM HOWITZER CARRIAGE M1

Parts to be Inspected in Order of Inspection	Points to Observe
The howitzer as a unit.	Note the general appearance. Test the smoothness of operation of the breech mechanism, both in opening and closing. Test the firing mechanism by firing two primers. Disassemble the breech mechanism and see that it is thoroughly cleaned and well lubricated (see item below on breechblock carrier). Note the condition of the bore for deposits on the lands and in the grooves, bruises on the gas check seat, erosion at the origin of rifling, and burs or roughness of the leveling plates. Note that the recoil slide is smooth, well lubricated, and free from rust or defects.
Breech recess and breechblock threads.	Note whether there are scores, bruises, or burs on the threads of the breechblock and breech recess.
Breechblock carrier and its attached parts.	While the breech mechanism is disassembled, note roughened or scored condition of the breechblock and driver bearing surfaces, firing mechanism housing, adapter, crosshead, crankshaft, and hinge pin. Test the operation of the safety latch and the operating lever latch; replace weak or broken springs.
Obturator spindle.	Note whether there is erosion of the venthole and primer chamber. Note the condition of the obturator spindle plug, and the threads on the end of the spindle. Look for a bruised split ring, torn gas check pad, broken or weakened obturator spindle spring. Try several primers in the obturator spindle plug; they

TM 9-331
61-62

INSPECTION AND ADJUSTMENT

Parts to be Inspected in Order of Inspection	Points to Observe
	should extend more than 1/8 inch when pressed in hard with the thumb or finger.
Firing Mechanism M1 and spare.	Disassemble and observe whether it is properly lubricated. Examine condition of firing pin. If the firing pin is weak or broken, replace. Observe the condition of the threads, the hole in the firing pin guide, and the bearing surfaces of the parts. Assemble the firing mechanism and seat a primer in the primer holder. Test the action of the spring.
Percussion hammer.	Test smoothness of operation. Note whether the rounded portion of the hammer face or the safety lug is worn or burred. Test the operation of the lockpin.
Counterbalance.	Test the operation of the mechanism and see that it functions properly at various degrees of elevation.

62. INSPECTION OF CARRIAGE.

The carriage as a unit.	Note the general appearance. Note whether the oil and grease fittings are clean and painted red, that a red ring has been painted around all oilholes, and that the carriage is painted in accordance with regulations. Note any gloss on the finish or any unnecessarily bright metal that would affect camouflage value.
Recoil mechanism.	Determine whether the replenisher and recuperator contain the proper amounts of reserve oil. Note whether the oil index and the replenisher piston function properly. Make sure breather

TM 9-331

155-MM HOWITZER M1 AND 155-MM HOWITZER CARRIAGE M1

Parts to be Inspected in Order of Inspection	Points to Observe
	holes in the rear of the replenisher piston guide are not plugged. Examine the mechanism for oil leaks. See that the wiper on the front of the recoil yoke tightly fits the recoil slide so as to fully exclude dirt.
Elevating mechanism.	Elevate and depress the weapon through the full extent of its travel. Note whether the mechanism operates without binding or interference and with no more than ½ turn backlash. Look for loose connections. See that the nut which retains the handwheel is tight. Check the function of the variable recoil mechanism. Examine the arc and pinion for undue wear, bruises, or burs. Check the lubrication.
Traversing mechanism.	Traverse the carriage throughout its complete movement. Note whether the operation is smooth. There should be no binding nor should the backlash be more than ½ handwheel turn. Look for loose mounting screws or nuts. Examine the arc and pinion for undue wear, bruises, or burs. Check the lubrication.
Top carriage.	Examine for breaks and cracks, and for loose attachment screws and nuts of the shields and the telescope mount. See that the pockets are clean and that the drain hole in each pocket is not closed.
Bottom carriage, trails, and spades.	Examine for breaks and cracks. Determine whether the trail hinge pins are properly lubricated. Examine accessory mountings on trails. Check the

INSPECTION AND ADJUSTMENT

Parts to be Inspected in Order of Inspection	Points to Observe
	operation of the trail lock; adjust if necessary. Look for undue wear on the lunette. Note the general condition of the spades and firing lock float.
Firing jack, handles, and traveling lock.	Note the condition of rack plunger; remove any rust. Check ease of operation. Examine ratchets. Determine whether traveling float is in place. Check condition and effectiveness of cover. Note security with which mounting holds the jack handles in traveling position. Examine the traveling lock and see that the chain on the traveling lock pin is sound.
Wheels and tires.	Check for any loose or missing nuts. Check the tires for correct pressure, 65 pounds per square inch. Note the condition of the tires and see whether the treads are taking the wear evenly. Raise and rotate the wheels to test for drag, bearing-adjustment, and wheel alinement.
Electric brakes.	Check current at each brake, ground connections, wiring for breaks in the insulation. Check plugs and socket for dirty or corroded blades or broken parts. Check the safety switch by pulling the safety switch lever and listening for the click caused by the armature contacting the magnet. Check the condition of the dry cell batteries and connections. Test the effectiveness of the brakes by making several stops. Test the action of the hand brakes.

TM 9-331
63

155-MM HOWITZER M1 AND 155-MM HOWITZER CARRIAGE M1

Section VIII

DISASSEMBLY AND ASSEMBLY

	Paragraph
General	63
Disassembly of Firing Mechanism M1	64
Assembly of Firing Mechanism M1	65
Disassembly of breech mechanism	66
Assembly of breech mechanism	67
Disassembly and assembly of operating lever latch	68
Disassembly and assembly of block rotating roller	69

63. GENERAL.

a. Cleaning, lubrication, and inspection, and incidents of use and breakage, make necessary the disassembly and assembly of the breech mechanism and various parts of the carriage. This work comes under two headings: that which can be performed by the battery personnel with the equipment furnished; and that which must be performed by ordnance maintenance personnel.

b. The battery personnel may, in general, do such dismounting as is required for the installation of the spare parts carried by the battery, and as is required for thorough inspection, cleaning, and lubrication. This work will be confined mainly to the breech mechanism and should be done in the manner prescribed and with proper tools.

c. Any difficulty which cannot be remedied by the prescribed methods will be brought to the attention of the ordnance maintenance personnel. Battery personnel will not attempt to dismount the howitzer from the cradle, or the cradle from the carriage, or to do any work on the recoil mechanism (other than prescribed in this manual) due to the impracticability of furnishing the organization with the necessary equipment for the performance of the work.

d. No filing of sighting equipment or howitzer parts will be done by the battery personnel except by order of the commanding officer.

e. The use of wrenches which do not fit snugly on the parts should be avoided. They not only will fail to tighten the parts properly, but they will damage the corners of nuts and screw heads. There is also danger of spreading the wrenches and rendering them useless.

f. A steel hammer should not be used directly on any part of the howitzer or machined part of the carriage. A copper or wooden mallet should be used, or a brass or copper drift, or a wooden block should be interposed between the blow and the part to be struck.

g. It is desirable to complete the subassembling of units before attempting the assembly of the units to the howitzer. In all assemblings,

DISASSEMBLY AND ASSEMBLY

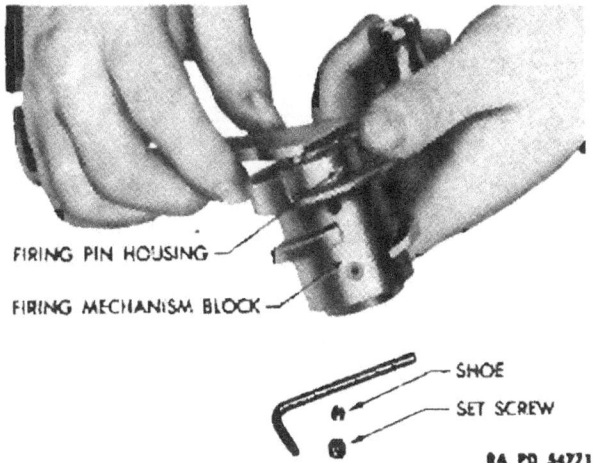

Figure 57—Removing the Firing Pin Housing

the bearings, sliding surfaces, threads, and finished parts should be cleaned and lubricated as directed.

64. DISASSEMBLING OF FIRING MECHANISM M1.

a. Remove both set screws from the firing mechanism block, using a ⅛-inch socket-head set screw wrench. Do not loosen the shoe under the set screw for the firing pin housing. Unscrew the firing pin housing from the rear end of the block, using the firing mechanism wrench (fig. 57).

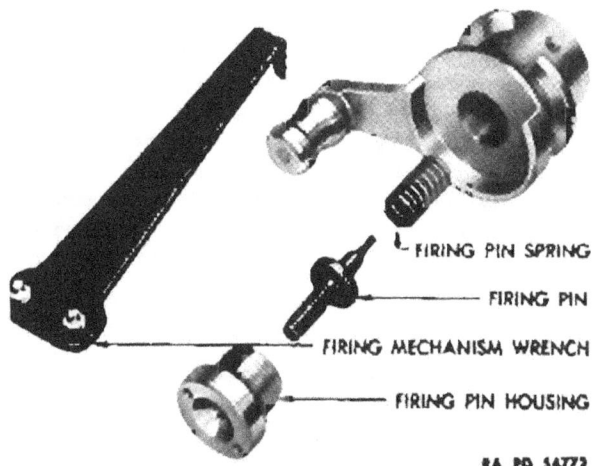

Figure 58—Parts in Rear of Firing Mechanism M1

TM 9-331
64-65

155-MM HOWITZER M1 AND 155-MM HOWITZER CARRIAGE M1

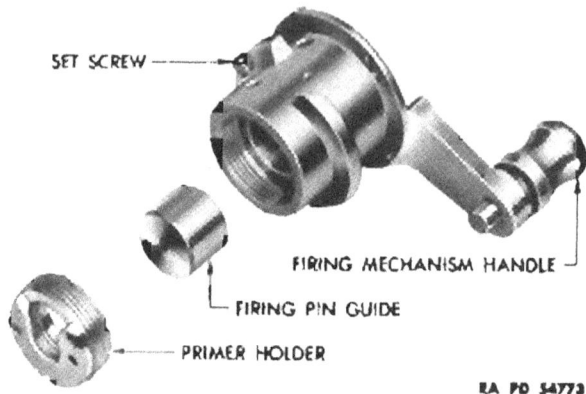

Figure 59 — Parts in Front of Firing Mechanism M1

b. Remove the firing pin and the firing pin spring from the rear of the block (fig. 58). Push the copper shoe out of the set screw hole, if it has not already come out.

c. Unscrew the primer holder (fig. 59) from the front of the block, using the firing mechanism wrench. The primer holder has left-hand threads and must be unscrewed to the right (clockwise) to be removed. Remove the firing pin guide.

d. The handle parts should not be removed from the arm of the block and the percussion hammer parts should not be removed from the adapter except to replace unserviceable parts or to correct malfunctioning. The pins used in assembling these parts are generally malformed in the course of disassembly, and the new parts must be peened or staked in place. It is desirable that these parts be replaced by experienced ordnance maintenance personnel, whenever possible.

e. Cleaning and Oiling. Clean all parts thoroughly with SOLVENT, dry-cleaning. Be sure all powder stains and traces of rust are removed, using CLOTH, crocus, if necessary. Wipe the surfaces dry and remove all particles of grit by using clean rags. Lubricate with a film of OIL, engine, lubricating, SAE 10, for temperatures below 32 F, and SAE 30 for temperatures above 32 F.

65. ASSEMBLY OF FIRING MECHANISM M1.

a. Place the firing pin guide in the forward end of the firing mechanism block with the hollow portion of the guide toward the rear of the block. Screw the primer holder into the forward end of the block. Use the firing mechanism wrench to seat the holder against the internal shoulder of the block, turning the holder to the left (counterclockwise) to install. Look

DISASSEMBLY AND ASSEMBLY

Figure 60 — Disconnecting the Counterbalance Mechanism

in the set screw hole and aline the nearest notch in the rear edge of the holder with the set screw hole. Insert and tighten the set screw, using the socket-head wrench.

b. Place the firing pin spring into the firing pin guide through the rear of the block. Insert the rounded end of the firing pin into the inner end of the firing pin housing, and screw the housing into the rear end of the block, using the firing mechanism wrench.

c. As the housing approaches its seat, it should be screwed carefully, making sure that the firing pin point properly enters its hole in the center of the guide. When the rear shoulder of the housing is firmly seated against the rear of the block, insert the copper shoe into the set screw hole to protect the threads of the firing pin housing. Screw in and tighten the set screw.

CAUTION: When the set screws are seated, they must be flush with or below the outside of the firing mechanism block.

66. DISASSEMBLY OF BREECH MECHANISM.

a. Cautions.

(1) Under no circumstances should an attempt be made to disassemble the breech mechanism with the breech in closed position. Failure to observe this rule may result in the displacement of the split rings and the gas check pad, and the dropping of the rings into the threads of the

TM 9-331
155-MM HOWITZER M1 AND 155-MM HOWITZER CARRIAGE M1

Figure 61 — Withdrawing the Control Arc

breech recess. This probably would cause serious damage to the rings and threads and prevent either swinging the breechblock out of the breech recess or returning of the breechblock to closed position.

(2) Should this condition occur, through accident or carelessness, no attempt should be made to force the breechblock or the carrier. With the breechblock in unlocked position, disconnect the counterbalance mechanism from the hinge pin (e, below) and remove the hinge pin (n, below). Several strong men then should carefully remove the complete carrier and breechblock assembly rearward from the weapon, leaving the obstructing parts in the recess to be removed after the carrier and breechblock assembly has been removed.

b. Remove the firing mechanism if it is installed in the weapon.

c. Open the breech and swing the carrier until a U-block can be placed between the shoulder of the tension rod eye and the counterbalance cylinder head (fig. 60). Then swing the carrier toward closed position until the tension rod eye can be lifted over the head of the pin on the hinge pin arm.

CAUTION: It is essential that the counterbalance mechanism be disconnected before disassembly of the breech mechanism is begun in order to protect the personnel and the materiel from the results of accidental closing of the breech.

d. With the breech in closed position, remove the control arc screw from the front face of the breech ring near the hinge. Withdraw the control arc (fig. 61) from its recess in the hinge lug. This makes the next operation possible.

e. Open the breech. With the carrier remaining in open position, lift the operating lever to latched position. This is possible due to the removal of the control arc. It places the breechblock in the position in which it

DISASSEMBLY AND ASSEMBLY

Figure 62—Placing Open Breech in "Closed" Position

would rest were the block sealed in the breech recess, and it permits the firing mechanism safety latch plunger to be retracted from the holes in the adapter and the firing mechanism housing (fig. 62) after the stop plate is removed. The latter condition is essential in the removing of these parts.

f. Remove the socket-head screw (fig. 63) which holds the safety latch plate to the rear face of the carrier. The plate and safety latch spring may fly off when the screw is removed and should be retained by the hand and removed after the screw is out. Move the safety latch to the right to the limit of its travel. Make certain that the safety latch plunger

Figure 63—Exploded View of Safety Latch Stop Plate

155-MM HOWITZER M1 AND 155-MM HOWITZER CARRIAGE M1

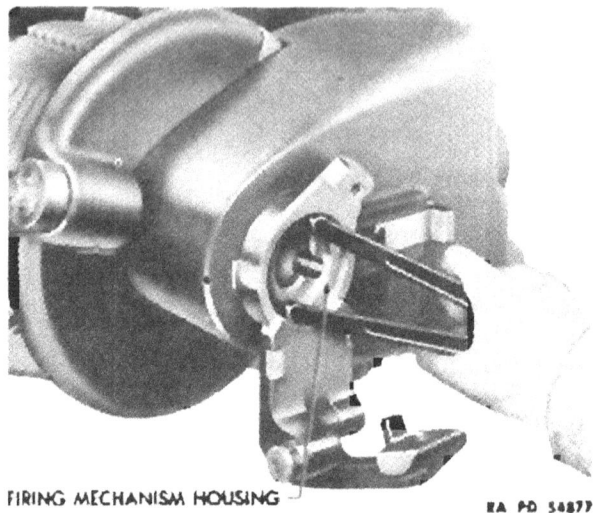

Figure 64 — Removing the Firing Mechanism Housing

does not contact the firing mechanism housing as it would score the outside of the housing when the latter was being unscrewed.

g. Unscrew the firing mechanism housing, using the firing mechanism housing wrench, an adjustable spanner, or the substitute tool described in the following note (fig. 64). This will release the adapter and the obturator assembly and spring.

NOTE: If the firing mechanism housing wrench is not available, a substitute tool may be made locally from a piece of steel plate 2½ inches wide, ¼ inch in thickness, and of convenient length. The working edge may be filed to the proper thickness. The housing may be rotated with a drift and hammer, but continuation of this practice tends to deform the edges of the slots and ruin the part.

h. Withdraw the adapter and the obturator spring from the rear of the carrier (fig. 65). Withdraw the obturator spindle from the front end of the breechblock, carrying with it the split rings, the gas check pad, the inner ring and the filling-in disk.

CAUTION: The obturator spindle plug, gasket, and vent bushing are not to be removed and replaced except by trained ordnance maintenance personnel.

i. Slide the firing mechanism safety latch and its plunger to the left, and withdraw it from the carrier (fig. 66). Withdraw the obturator spindle bearing sleeve rearward out of the breechblock and carrier.

TM 9-331
66

DISASSEMBLY AND ASSEMBLY

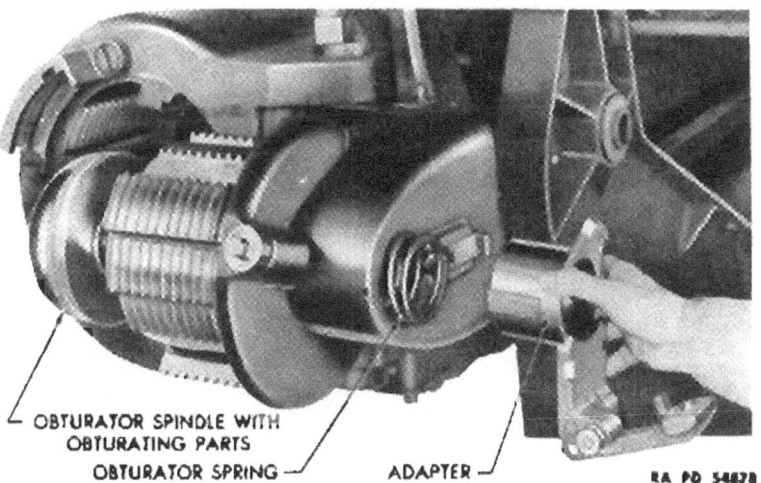

Figure 65—Removing Adapter, and Obturator Assembly and Spring

j. Slide the breechblock forward off the driver (fig. 67), rotating the breechblock by slightly manipulating the operating lever, if necessary, to free the cam roller from the cam slot. The driver should be held while the block is being removed.

k. Remove the locking screw which locks the retaining ring to the carrier at the forward end of the driver hub. Unscrew the retaining ring from the carrier, using the retaining ring wrench or an adjustable spanner (fig. 68). Slide the breechblock driver forward off the carrier.

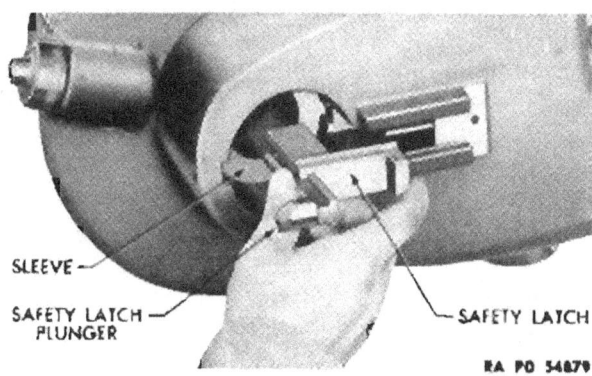

Figure 66—Removing the Safety Latch and Plunger

TM 9-331

155-MM HOWITZER M1 AND 155-MM HOWITZER CARRIAGE M1

Figure 67 — Removing the Breechblock

l. Remove the lever retaining screw (fig. 69) from the hub of the operating lever. Withdraw the lever from its slot in the crankshaft bushing. Unscrew and remove the hexagon-headed bushing retaining screw from the rear of the carrier near the right side. This screw is threaded only near its head, and its long shank engages a groove encircling the bushing, retaining the bushing in its seat in the carrier.

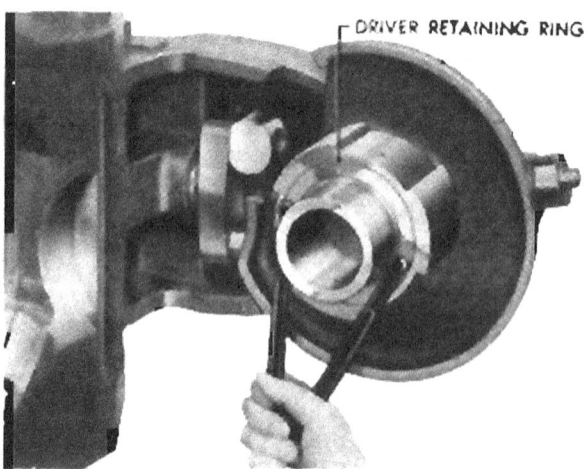

Figure 68 — Removing the Driver Retaining Ring

DISASSEMBLY AND ASSEMBLY

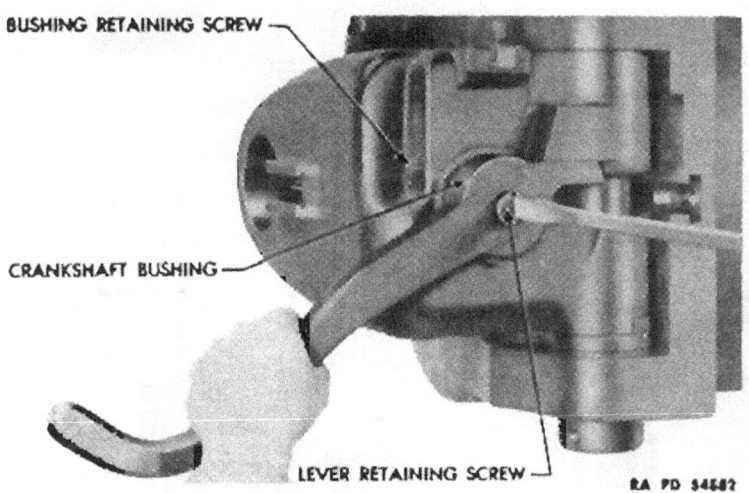

Figure 69 — Removing the Operating Lever

m. Slide the crosshead off its pivot on the arm of the crankshaft. The cam roller on the other arm of the crankshaft is retained in a semi-permanent manner, and the roller should not be removed except when

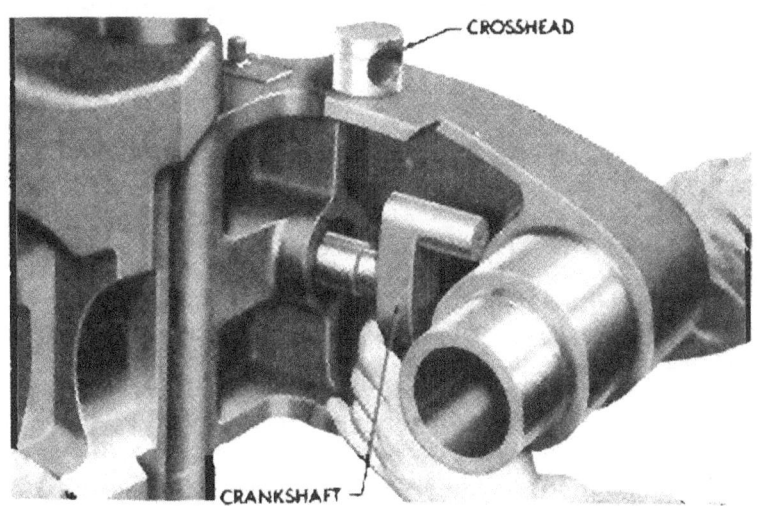

Figure 70 — Removing the Crankshaft

TM 9-331
66

155-MM HOWITZER M1 AND 155-MM HOWITZER CARRIAGE M1

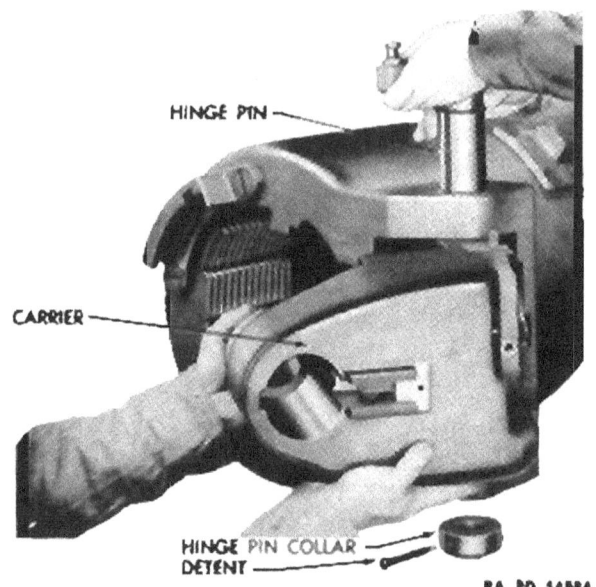

Figure 71 — Removing the Carrier Hinge Pin

replacement is necessary. With the right hand, withdraw the crankshaft bushing (fig. 72) from its bore in the right end of the carrier while, with the left hand, remove the crankshaft through the inside of the carrier (fig. 70).

n. Drive the detent out of the collar on the lower end of the carrier hinge pin. Remove the collar. While the carrier is being held, withdraw the hinge pin from the hinge pin lug (fig. 71). A slight movement of the carrier may assist in the removal of the pin. Remove the carrier from the hinge pin lug. The hinge pin driving washer generally will adhere to the carrier and should not be permitted to fall. Remove the bearing disk from the lower hinge pin lug.

CAUTION: The counterbalance mechanism is not to be disassembled except to correct malfunction or to replace damaged parts, and such work is only to be done by ordnance maintenance personnel.

a. Cleaning and Oiling.

(1) Clean all parts thoroughly with SOLVENT, dry-cleaning. Be sure all powder stains and traces of rust are removed, using CLOTH, crocus, if necessary. Wipe the surfaces dry and remove all particles of grit by using clean rags. Lubricate with a film of OIL, engine, lubricating, SAE 10, for temperatures below 32 F, and SAE 30 for temperatures above 32 F.

DISASSEMBLY AND ASSEMBLY

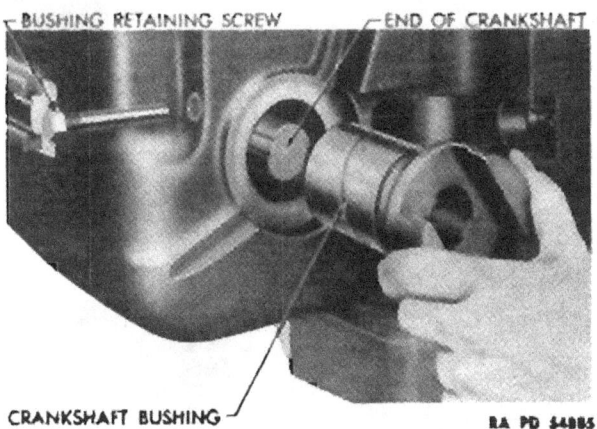

Figure 72—Inserting the Crankshaft Bushing

(2) Apply OIL, engine, lubricating, SAE 10, to the gas check pad and rub it in well. This is to keep the pad compressible. A pad of the proper resiliency will yield slightly under heavy pressure from the thumbs.

67. ASSEMBLY OF BREECH MECHANISM.

a. Place the bearing disk on the lower hinge lug of the breech ring, fitting the small hole in the disk over the pin inserted in the lug. Assemble the hinge pin driving washer to the lower hinge portion of the carrier, and place the carrier in its position between the hinge lugs and on the bearing disk.

b. While the carrier is being supported, insert the hinge pin downward through the hinge pin lug and carrier. Slowly rotate the hinge pin so that the squared portion of the pin will enter the square hole in the driving washer, with the radial arm of the pin pointing toward the left end of the carrier. Place the collar on the lower end of the hinge pin, alining the holes in the collar and pin. Insert the detent through the holes to secure the collar to the pin.

c. From the inside of the carrier, insert the crankshaft into its bore in the right side of the carrier. Support the crank end of the crankshaft inside the carrier, with the crosshead arm of the crankshaft uppermost. Start the small end of the crankshaft bushing onto the outer end of the crankshaft from the outside of the carrier (fig. 72) and guide the bushing into the bore of the carrier.

d. Push the bushing home and screw the bushing retaining screw (fig. 72) into the carrier to secure the bushing. Insert the operating

TM 9-331
67

155-MM HOWITZER M1 AND 155-MM HOWITZER CARRIAGE M1

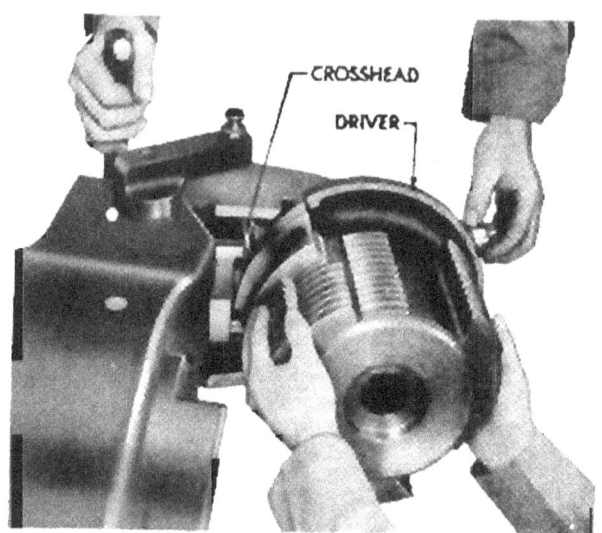

Figure 73 – Installing the Breechblock

lever into the crankshaft bushing, handle upward, and secure the lever to the crankshaft with the lever retaining screw. Place the crosshead on its pivot on the arm of the crankshaft.

e. Slide the driver, flange rearward, onto its bearing on the front of the carrier. Screw the retaining ring to the carrier to retain the driver. Lock the retaining ring in place with its locking screw. Position the crosshead horizontally near the top of its travel.

f. Start the breechblock onto the carrier, alining the breechblock so

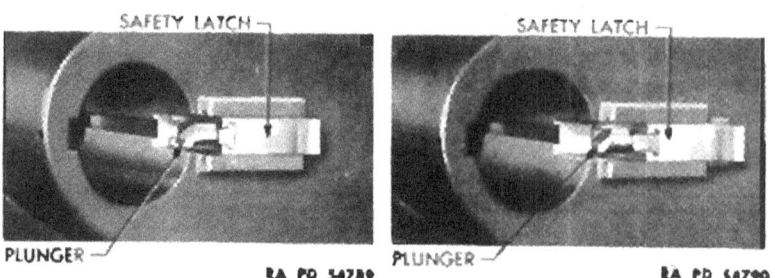

Figure 74 – Safety Latch Plunger Flush with Front End of Latch

Figure 75 – Safety Latch Plunger Clear of Bore in the Carrier

112

TM 9-331
67

DISASSEMBLY AND ASSEMBLY

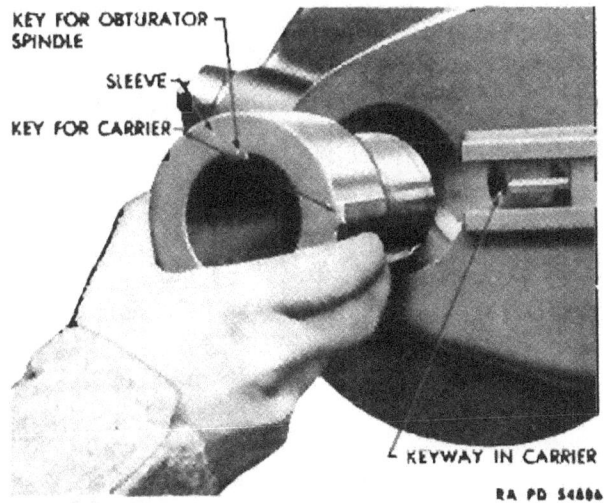

Figure 76—Inserting the Obturator Spindle Bearing Sleeve

that the crosshead can be guided into its groove in the breechblock while, at the same time, holding the driver in such a position that the cam lug on the breechblock will enter the notch in the flange of the driver (fig. 73). With the crosshead entered into its groove, slide the breechblock rearward onto the driver. Latch the operating lever.

g. Start the safety latch into the rear lug of the carrier from the left. Insert the safety latch plunger into the safety latch. Push the plunger forward until its front end is approximately flush with the front end of the latch (fig. 74). Then press the latch toward the right, moving the plunger slightly forward and backward until the leftward-projecting pin in the safety latch recess of the carrier enters the hole in the right side of the body of the plunger. Push the latch and plunger to the right as far as they will go (fig. 75). The plunger will now clear the bore of the carrier and permit assembly of the adapter.

CAUTION: Under no circumstances should the breechblock be swung into the breech recess until the obturator spindle, pad, rings, and filling-in disk have been properly secured in position by inserting the sleeve, obturator spring, and adapter, and screwing the firing mechanism housing to the correct assembled position on the obturator spindle (par. 66 a (1) and (2)).

h. Insert the obturator spindle bearing sleeve, smaller end first, into the rear of the carrier bore (fig. 76). Fit the key on the right side of the sleeve into the keyway in the bore, and push the sleeve forward as far as it will go.

113

TM 9-331
67

155-MM HOWITZER M1 AND 155-MM HOWITZER CARRIAGE M1

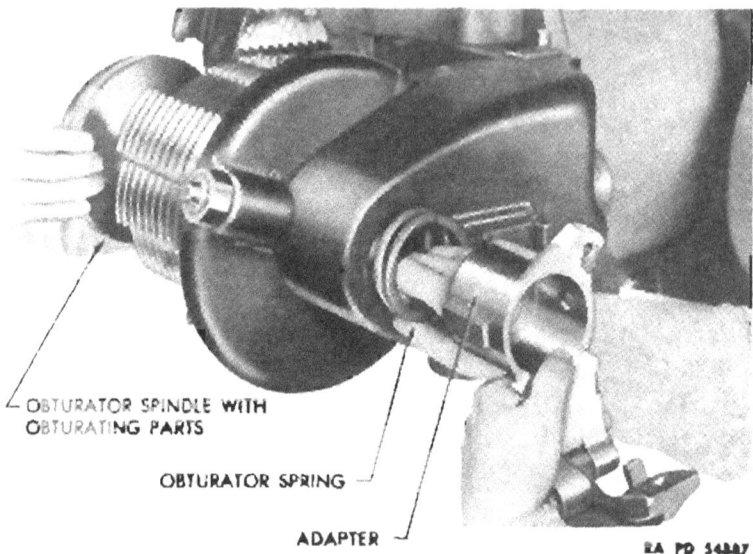

Figure 77 — Inserting Obturator Spring, Adapter, and Obturator Spindle

i. Place the front split ring (smaller diameter), gas check pad, rear split ring, inner ring, and filling-in disk, in the order named, on the rear of the head of the obturator spindle. Insert the spindle into the bore of the breechblock and through the bore of the sleeve, fitting the open keyway near the end of the spindle on the key in the upper rear end of the bore of the sleeve.

j. Place the obturator spindle spring into the rear bore of the carrier and press the adapter in behind the spring, fitting the key on the left side of the adapter into the keyway in the left wall of the carrier bore. Insert the firing mechanism housing into the adapter and while pressing forward on the adapter to compress the spring, screw the housing onto the rear end of the obturator spindle. Check the split rings for proper position under the obturator spindle head and screw the housing until it seats. Back off the housing slightly, if necessary, to bring the slot in its right wall into alinement with the hole in the adapter and with the safety latch plunger.

k. Push the safety latch to the left, inserting the pointed end of the plunger in the holes in the adapter and housing. Insert the spring into the hole in the right end of the safety latch and assemble the stop plate to the carrier with its retaining socket-head screw. Test the safety latch against its stop.

DISASSEMBLY AND ASSEMBLY

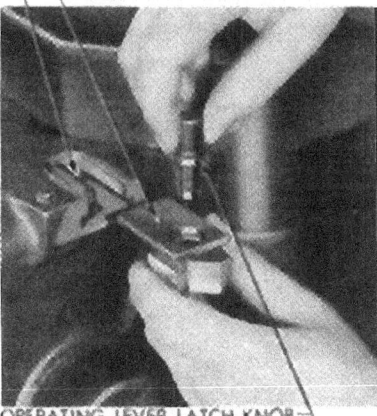

Figure 78 — Disassembled View of Operating Lever Latch

l. Lower the operating lever against its open position stop on the carrier and carefully swing the carrier to closed position, easing the breechblock into the breech recess without permitting the edges of one part to strike those of another. Insert the control arc into its slot in the hinge lug of the breech ring. Secure the arc with the control arc screw.

m. Swing open the carrier slightly until the slotted eye on the end of the counterbalance tension rod can be placed over the head of the pin on the arm of the carrier hinge pin. Swing open the carrier sufficiently to permit the U-block to be removed. Close the breech and lubricate through the fittings as indicated on the lubrication guide (fig. 50).

68. DISASSEMBLY AND ASSEMBLY OF OPERATING LEVER LATCH.

a. Disassembly. Unscrew the knob of the operating lever latch to the left (counterclockwise). When the knob is withdrawn, the latch will slide outwardly to the right under the tension of its spring (fig. 78). Clean and lubricate.

b. Assembly. Insert the spring in its bore in the back of the latch recess. Slide the latch into its groove and hold it against the tension of the spring until the knob can be screwed home.

69. DISASSEMBLY AND ASSEMBLY OF BLOCK ROTATING ROLLER.

a. Disassembly. Remove the cotter pin from the block rotating roller lug on the rear face of the driver. Withdraw the block rotating roller pin and release the block rotating roller (fig. 79).

155-MM HOWITZER M1 AND 155-MM HOWITZER CARRIAGE M1

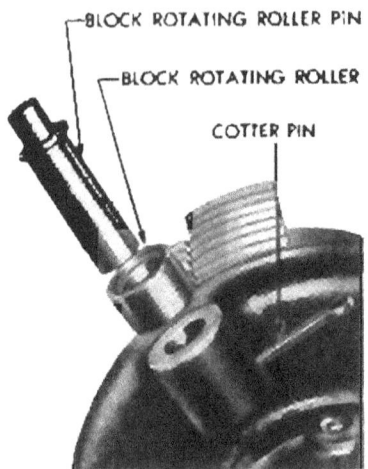

Figure 79—Exploded View of Block Rotating Roller

b. **Assembly.** Slide the block rotating roller on its pin and insert the pin in the bore of the block rotating roller lug on the driver. Aline the holes in the pin and lug and insert the cotter pin.

CHAPTER 3
SIGHTING AND FIRE-CONTROL EQUIPMENT

Section I
SIGHTING EQUIPMENT

	Paragraph
General	70
Care and preservation	71
Telescope mount M25, with panoramic telescope M12	72
Aiming post M1	73
Aiming post light M14	74
Gunner's quadrant M1, or M1918	75
Bore sight	76
Testing target	77
Procedure for bore sighting	78

70. GENERAL.

a. The laying equipment for the 155-mm Howitzer M1 consists of the Telescope Mount M25 with the Panoramic Telescope M12. The equipment is designed to lay the howitzer in azimuth and elevation for either direct or indirect fire. The mount is secured to the left side of the howitzer carriage and is in line with the axis of trunnions.

b. The Aiming Post M1 is used as an aiming point and for checking lateral deflection of the carriage in indirect fire. The Aiming Post Light M14 is used on the aiming post for night firing.

c. The gunner's quadrant is used for determining elevation of the howitzer and for checking the alinement in elevation of the barrel bore with respect to the telescope mount. Either the Gunner's Quadrant M1, or the Gunner's Quadrant M1918 may be furnished.

d. The bore sight and testing target are used during the bore sighting operation for verification and alinement of sights.

71. CARE AND PRESERVATION.

a. General.

(1) The instructions given here supplement the instructions pertaining to individual items of sighting and fire-control equipment described in this and the following section.

(2) Fire-control and sighting instruments are, in general, rugged and suited for the purpose for which they have been designed. They will not, however, stand rough handling, or abuse. Inaccuracy or malfunctioning will result from mistreatment.

TM 9-331
71

155-MM HOWITZER M1 AND 155-MM HOWITZER CARRIAGE M1

(3) Disassembly and assembly by the using arms are permitted only to the extent for which tools and parts have been provided. Other replacements and repairs are the responsibility of other maintenance personnel, but may be performed by the using arm personnel, when circumstances permit, within the discretion of the pertinent ordnance officer.

(4) Keep the instruments as dry as possible. If the instrument is wet, dry it carefully before placing it into its carrying case.

(5) When not in use, keep the instruments in the carrying case provided, or in the condition indicated for traveling.

(6) Any instruments which indicate incorrectly, or fail to function properly, after the authorized tests and adjustments have been made, are to be turned in for repair to ordnance personnel. Adjustments other than those for which tools have been provided are not to be performed by the using arms unless directed by the pertinent ordnance officer.

(7) No painting of fire-control equipment by the using arms is permitted.

(8) Many worm drives have throwout mechanisms to permit rapid motion through large angles. When using these mechanisms, it is essential that the throwout lever be fully depressed to prevent injury to the worm and gear teeth.

(9) When using a tripod with adjustable legs, be certain that the legs are clamped tightly to prevent possibility of collapse.

(10) When setting up tripods on sloping terrain, place 2 legs on the downhill side to provide maximum stability.

(11) Dry cell batteries should always be removed from the battery cases when not in use, in order that their deterioration on long standing will not damage the light. When not in use, the various parts of the lights should be kept in the chest provided.

b. Leather Articles. Care and preservation of leather articles are covered in TM 9-850.

c. Optical Parts.

(1) To obtain satisfactory vision, it is necessary that the exposed surfaces of the lenses and other parts be kept clean and dry. Corrosion and etching of the surface of the glass, which greatly interfere with the good optical qualities of the instrument, can be prevented, or greatly retarded, by keeping the glass clean and dry.

(2) Under no condition will polishing liquids, pastes, or abrasives be used for polishing lenses and windows.

(3) For wiping optical parts, use only PAPER, lens, tissue. Use of cleaning cloths in the field is not permitted. To remove dust, brush the glass lightly with a clean camel's-hair brush and rap the brush against a hard body in order to knock out the small particles of dust that cling to the hairs. Repeat this operation until all dust is removed. With some

SIGHTING EQUIPMENT

instruments, an additional brush with coarse bristles is provided for cleaning mechanical parts; it is essential that each brush be used only for the purpose intended.

(4) Exercise particular care to keep optical parts free from oil and grease. Do not wipe the lenses or windows with the fingers. To remove oil or grease from optical surfaces, apply ALCOHOL, ethyl, or SOAP, liquid, lens cleaning, with a tuft of clean lens paper and rub the surface gently with clean lens paper. If alcohol is not available, breathe heavily on the glass and wipe it off with clean lens paper; repeat this operation several times until the glass is clean.

(5) Moisture may condense on the optical parts of the instrument when the temperature of the parts is lower than that of the surrounding air. This moisture, if not excessive, can be removed by placing the instrument in a warm place. Heat from strongly concentrated sources should not be applied directly, as it may cause unequal expansion of parts thereby resulting in damage of optical parts and inaccuracies of observation.

d. Lubricants.

(1) Where lubrication with oil is indicated, use OIL, lubricating, for aircraft instruments and machine guns.

(2) Where lubrication with grease is indicated use GREASE, lubricating, special.

72. TELESCOPE MOUNT M25 WITH PANORAMIC TELESCOPE M12.

a. Description.

(1) The Telescope Mount M25 and the Panoramic Telescope M12 (figs. 80 and 81) are mounted on the left side of the 155-mm Howitzer Carriage M1 and form the sighting equipment for laying the howitzer in azimuth and elevation.

(2) The telescope mount (fig. 82) contains longitudinal and cross-leveling mechanisms. The cross-level is habitually kept centered by operating the cross-leveling knob. The centering of the cross-level automatically introduces an azimuth correction which compensates for any error produced when the howitzer is elevated with the trunnions out of level. The longitudinal level is operated by the elevating knob. The centering of the longitudinal level places the azimuth scale of the telescope adapter in a true horizontal plane, so that azimuth angles can be set accurately.

(3) The Panoramic Telescope M12 (fig. 83) is a conventional panoramic telescope with eyepiece offset 45 degrees to permit the observer to stand clear of the howitzer. It has cylindrical locating surfaces at the middle and bottom for centering the telescope in its socket. The line of sight is elevated or depressed by the knob at the top, and coarse and fine

TM 9-331
72

155-MM HOWITZER M1 AND 155-MM HOWITZER CARRIAGE M1

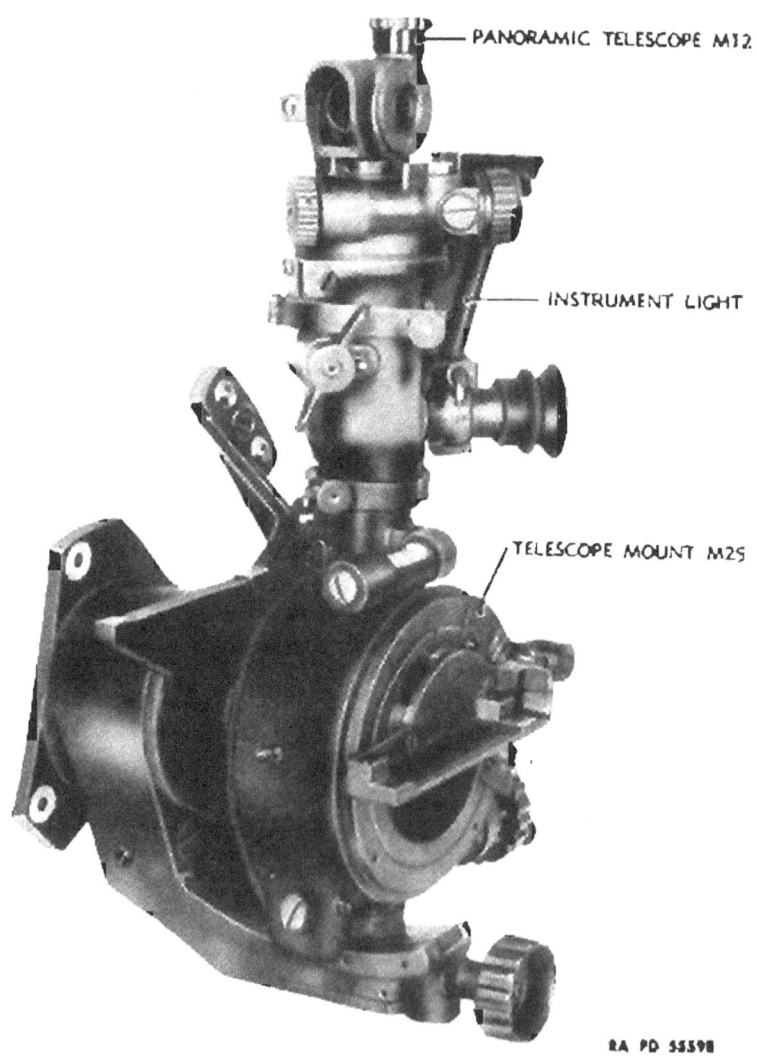

Figure 80 — Arrangement of Sighting Equipment — Left Side

SIGHTING EQUIPMENT

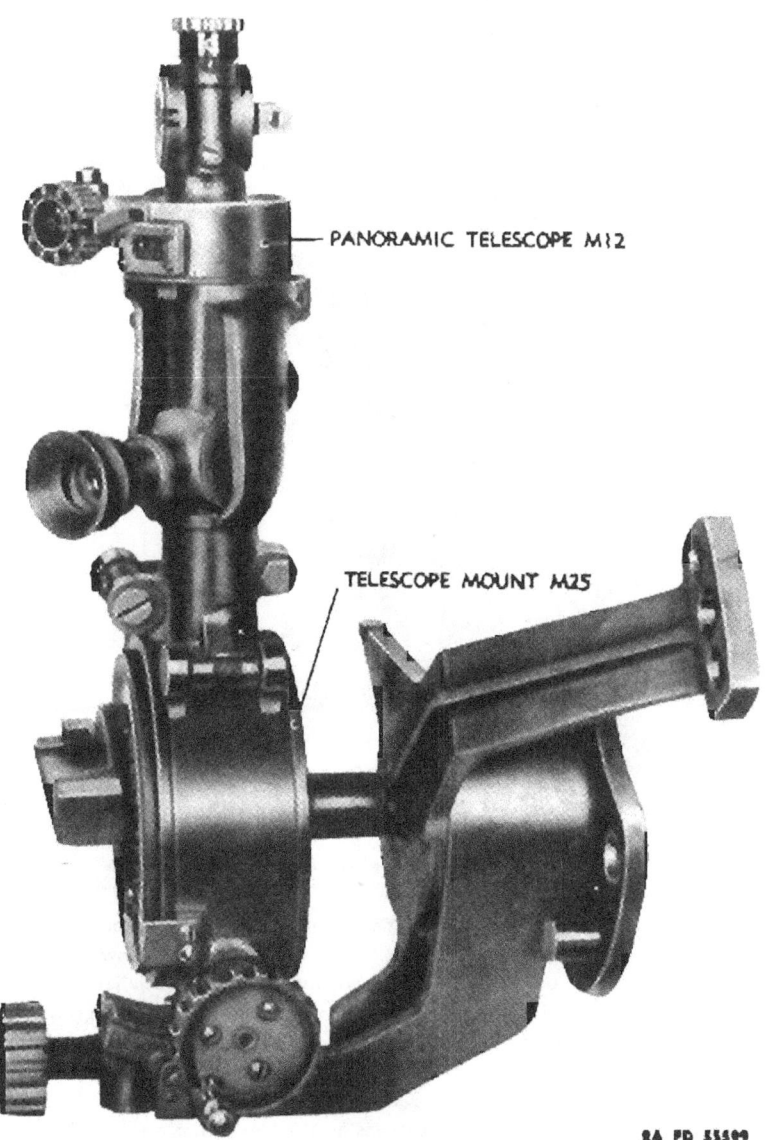

Figure 81—Arrangement of Sighting Equipment—Rear

155-MM HOWITZER M1 AND 155-MM HOWITZER CARRIAGE M1

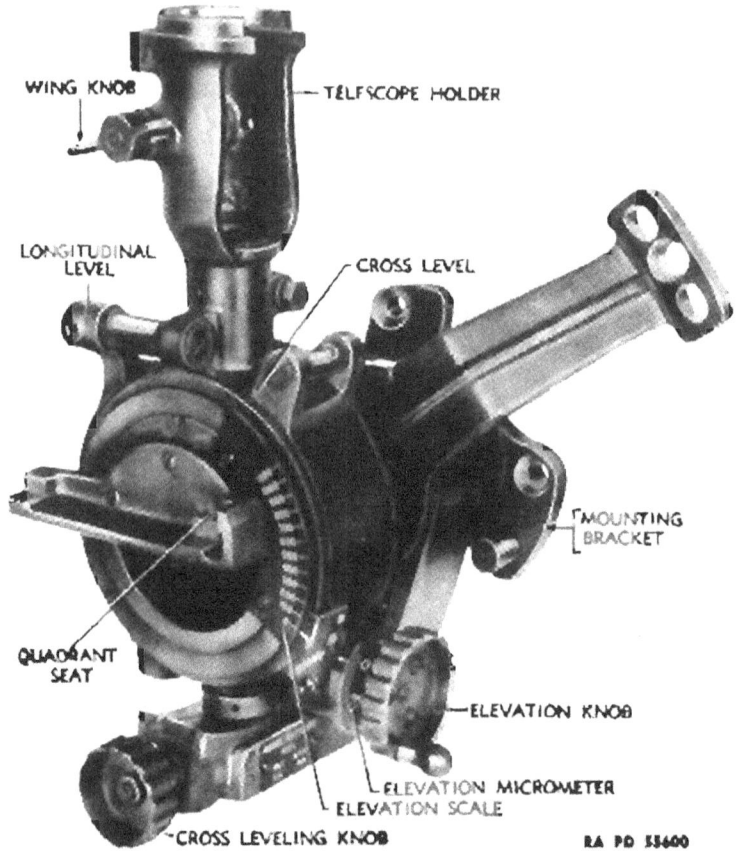

Figure 82 — Telescope Mount M25

index graduations are provided for the normal (zero elevation) position. The head is moved in azimuth by the azimuth worm knob which has a throwout lever to permit disengagement for rapid motion when required. The recticle (fig. 84) includes crosslines and a deflection scale graduated in five-mil intervals and numbered at 50-mil intervals. The azimuth scale is graduated in 100-mil intervals, numbered progressively from 0 to 32 in 2 consecutive semicircles. Azimuth micrometer indications (1-mil intervals) supplement the indications on the azimuth scale.

(4) The Instrument Light M16 is provided for lighting Telescope Mount M25 and Panoramic Telescope M12. The light clamps to the body of the panoramic telescope. It consists of a case for holding 2 dry

SIGHTING EQUIPMENT

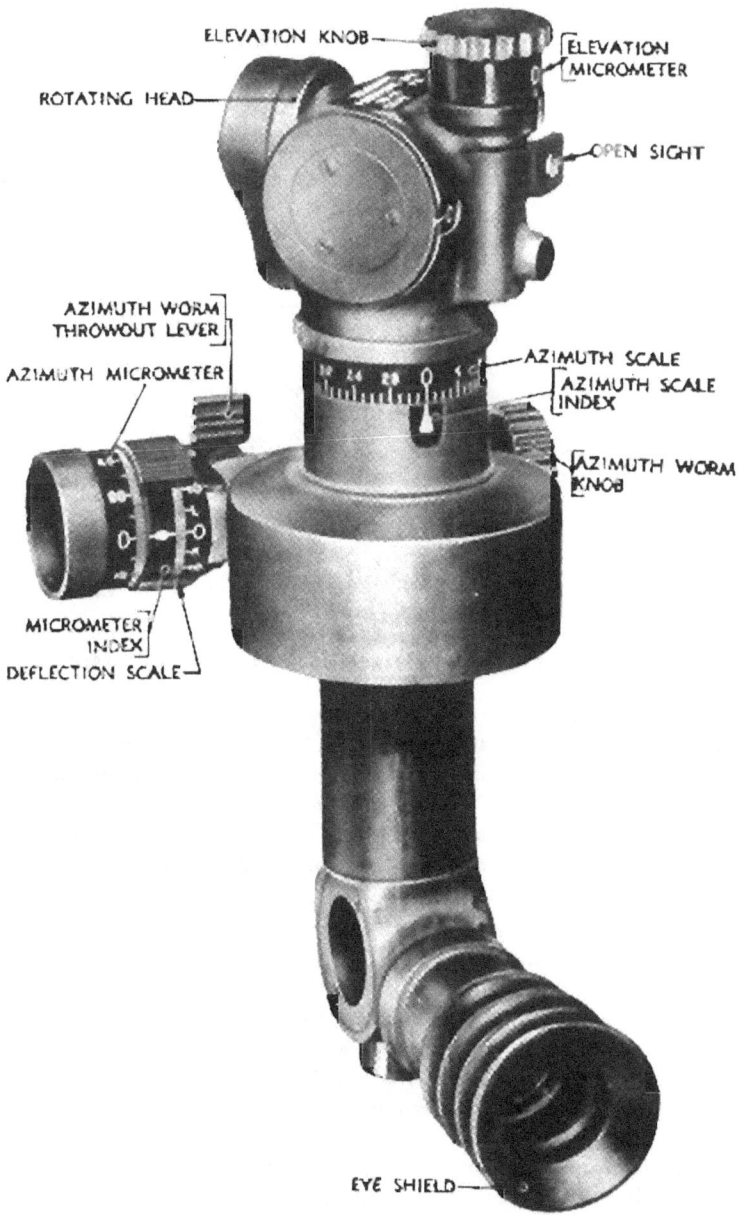

Figure 83 — Panoramic Telescope M12

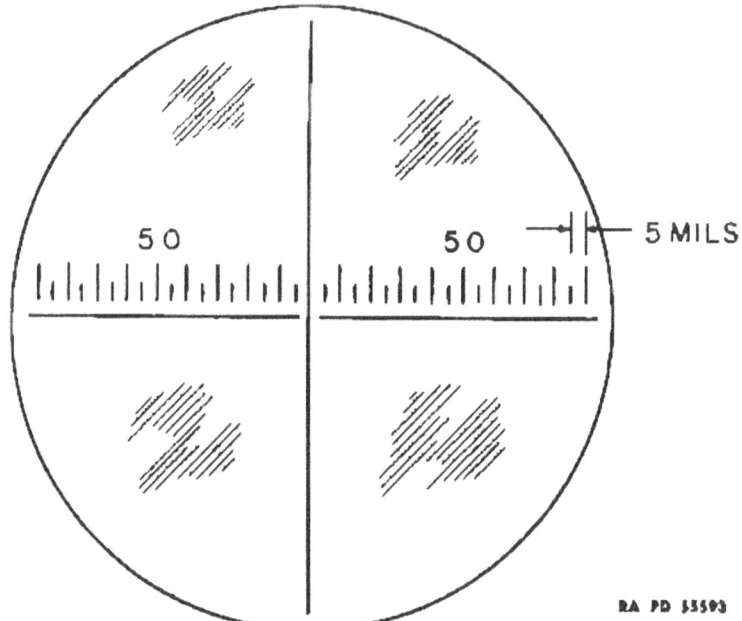

Figure 84 — Panoramic Telescope M12 — Reticle Pattern

cell batteries, clamps for mounting, 2 brackets which support 2 plastic rods, a finger light and cable, and a switch. The plastic rods carry light from a single lamp to both the reticle and the azimuth micrometer of the panoramic telescope. The finger light is used for illuminating the elevation scale and micrometer and for checking the level bubbles on the mount. The switch, protected by a shield, controls the lamps.

b. **Operation.**

(1) The Telescope Mount M25 and Panoramic Telescope M12 may be used in direct or indirect laying.

(2) To place the telescope in its socket, remove the telescope from the carrying case, turn the wing knob on the socket to its extreme counter-clockwise position and place the telescope gently in the socket. Exert slight downward pressure to insure that the telescope is properly seated. Release the wing knob. Uncover both levels of the telescope mount.

(3) DIRECT AND INDIRECT LAYING FOR ELEVATION. Three methods are available for setting the desired quadrant elevation into the telescope mount. It is essential that the mount be kept cross-leveled at all times during the operation. If the mount is not cross-leveled an erroneous azimuth will be applied to the howitzer as it is elevated or depressed. The procedure for indirect laying is the same as for direct laying.

SIGHTING EQUIPMENT

(a) Turn the elevation knob until the necessary quadrant elevation is indicated on the elevation scale and micrometer. This operation depresses the mount by an angular displacement equal to the necessary howitzer elevation. Rotate the elevation handwheel of the howitzer carriage until the longitudinal level bubble is centered and the howitzer is then laid in elevation.

(b) The angle of site and angle of elevation can be added algebraically by setting the elevation angle, as obtained from the firing table for the range, on the elevation scale and micrometer, and then elevating the howitzer until the intersection of the crosslines in the telescope fall on the target. When laying by this method, it is imperative that the elevation micrometer on the telescope be set at "ZERO."

(c) When a high degree of accuracy is desired, proceed as in subparagraph b (3) (a) above and then place an accurately-set gunner's quadrant on the quadrant seat of the bracket. Rotate the elevation handwheel of the howitzer carriage until the level bubble of the quadrant is centered.

(4) DIRECT LAYING FOR AZIMUTH. Set off the required deflection correction on the azimuth micrometer of the telescope. Bring the reticle crosslines of the reticle to bear on the target by turning the traversing handwheel of the howitzer carriage. It may be necessary to rotate the elevation knob on the telescope to bring the target into the field of view but this is not permitted when the procedure given in subparagraph b (3) (b) above is followed.

(5) INDIRECT LAYING FOR AZIMUTH. Set off the required deflection correction on the azimuth micrometer of the telescope. Set off the azimuth of the target with respect to the aiming point on the azimuth scale and micrometer of the panoramic telescope. Traverse the howitzer carriage by turning the traversing handwheel until the vertical crossline of the telescope reticle falls on the aiming point. It may be necessary to rotate the elevation knob on the telescope to bring the aiming point into the field of view.

c. **Preparation for Travel.** Turn the wing knob counterclockwise and lift out the panormic telescope. Place the telescope in the carrying case. Protect both levels by closing their covers.

d. **Test and Adjustment.**

(1) The mechanisms of the sighting and laying equipment should be tested periodically for lost motion. Lost motion of the panoramic telescope in its socket (tangent adjusting screws too loose) is indicated, if the telescope shifts in its socket under a light twisting pressure. Lost motion in the azimuth compensating mechanism of Telescope Mount M25 can be felt as a freedom or shake of the telescope socket. Lost motion in the micrometer mechanisms can be detected by operating the

155-MM HOWITZER M1 AND 155-MM HOWITZER CARRIAGE M1

mechanism first in one direction and then in the opposite direction, returning both times to the same micrometer setting. If the telescope or level vial which is operated by the mechanism does not return to the same aiming point or level position, it is an indication of lost motion in the mechanism.

(2) The effect of small amounts of lost motion can be eliminated by habitually making the last movement always in the same direction. The last movement in setting and laying for deflection should be from left to right. The last movement in setting the scales should be in the direction of increasing the reading. Lost motion in the elevating mechanism of the carriage can be taken up by moving the elevating handwheel against the greatest resistance.

(3) If an appreciable degree of lost motion exists in the sighting equipment, adjustment should be made without delay by ordnance personnel. Qualified battery personnel are authorized to correct looseness of the panoramic telescope in the socket of the mount by an adjustment of the tangent adjusting screws. Care must be exercised that there is no undue pressure by either tangent screws. The panoramic telescope should seat firmly without binding.

(4) Bore sighting procedure for verification and alinement of telescope mount and telescope is described in paragraph 78.

c. Care and Preservation.

(1) Refer to paragraph 71 for general care and preservation instructions pertaining to this equipment.

(2) The lubrication fittings on the telescope mount are not to be used by battery personnel. Do not apply lubricant through these fittings.

(3) Stops are provided to limit the longitudinal and cross-leveling motions. No attempt should be made to force the mechanisms beyond the stops.

(4) Keep the level vials covered at all times when not in use.

(5) Disassembling of the equipment, other than such disassembling as is incident to normal operating procedure, is not permitted.

(6) When using the throwout lever of the panoramic telescope, push the lever all the way against its stop. If the lever is pushed only part of the way, the gear teeth inside the telescope will scrape against each other as the head is rotated and will wear rapidly.

73. AIMING POST M1.

a. Two of these aiming posts are furnished with each 155-mm Howitzer Carriage M1. Each aiming post (fig. 85) consists of 2 tubular sections, each section approximately 4 feet long. The lower section has a metal point for embedding the post in the ground, and both sections are provided with halves of a joint-and-catch fitting. The parts are painted with alternate 4-inch red and white bands. A canvas cover, holding both

TM 9-331
73-74

SIGHTING EQUIPMENT

Figure 85—Aiming Post M1

sections, is provided. Should it be necessary to drive the lower section into the ground, interpose a wood block or use other means to insure that the surface mating with the upper part will not be injured.

b. One aiming post may be used as an aiming point. Two are used, set up in line with the telescope to check azimuth deflection caused by shifting of the carriage in firing.

74. AIMING POST LIGHT M14.

a. The Aiming Post Light M14 is a device for illuminating aiming posts for night firing.

b. The Aiming Post Light M14 (fig. 86) consists of a battery case for 2 BA-30 batteries with a lamp housing and a toggle switch. A metal hood is provided for the lamp. When the hood is not in use, it is carried around the battery case. Illumination is furnished by a 3-volt lamp. A reflector is mounted in the back of the lamp and a color filter can be attached to the front of the lamp housing (fig. 87).

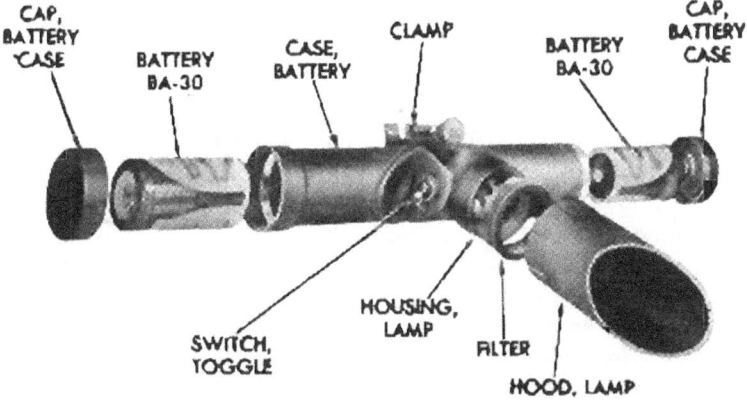

Figure 86—Aiming Post Light M14—Assembled and Mounted on Post

TM 9-331
74-75

155-MM HOWITZER M1 AND 155-MM HOWITZER CARRIAGE M1

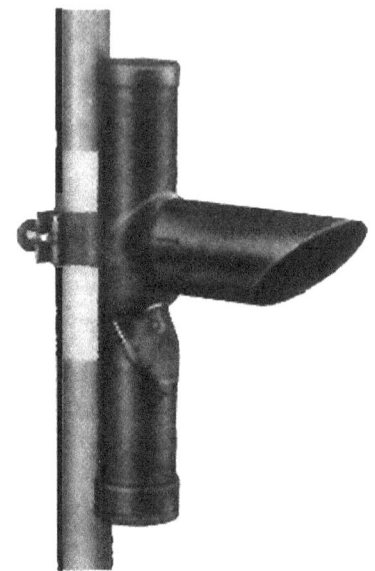

RA PD 49877

Figure 87 — Aiming Post Light M14 — Exploded View

c. A clamp is provided to secure the lamp to the aiming post.

d. A chest is provided to carry a section set, comprising 2 aiming post lights, (1 with red, 1 with green filters); 8 BA-30 batteries, and 2 spare lamps.

e. The batteries should always be removed from the battery case when not in use in order that their deterioration on long standing will not damage the light. When not in use, the various parts of the light should be kept in the chest provided.

75. GUNNER'S QUADRANT M1 OR M1918.

a. *General.* The gunner's quadrant is used for measuring the elevation or depression of the piece. Either the Gunner's Quadrant M1 (fig. 88) or the Gunner's Quadrant M1918 (fig. 89) may be furnished.

b. *Description.* The quadrant has a sector-shaped frame to which is pivoted an arm carrying a level. The frame has 2 scales with corresponding reference surfaces. One scale and reference surface are used for elevations from 0 to 800 mils, and the other for elevations from 800 to 1,600 mils. Notches on the frame, which engage a plunger in the arm, permit rapid setting of the arm in 10-mil steps. The level has a 10-mil smooth motion at each step to provide a fine indication supplementing the coarse scale reading. The method of obtaining the fine indication is

SIGHTING EQUIPMENT

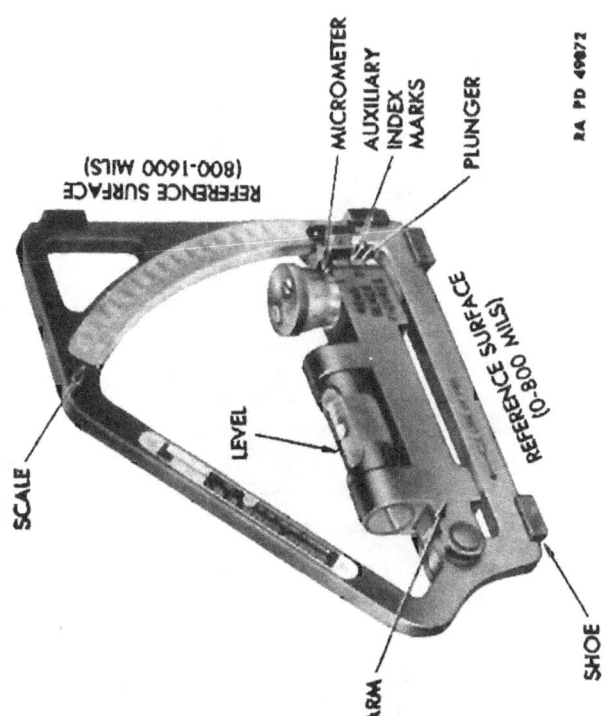

Figure 88 — Gunner's Quadrant M1

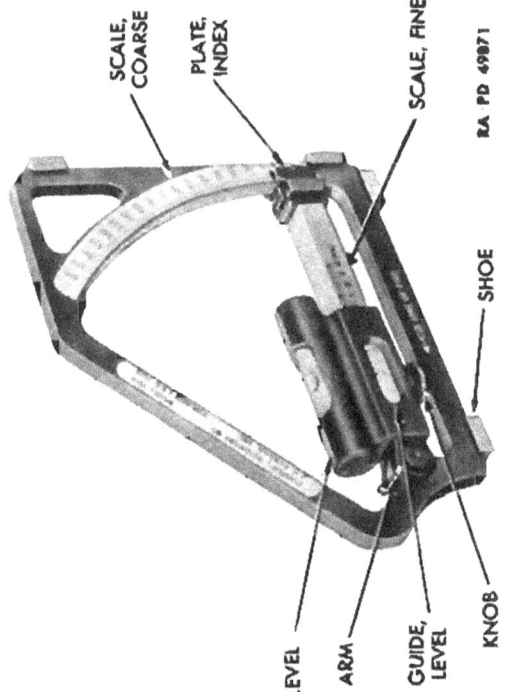

Figure 89 — Gunner's Quadrant M1918

SIGHTING EQUIPMENT

the principal difference between the 2 quadrants. The M1918 has a slightly curved arm with the level guide arranged to slide along the arm; the M1 has a micrometer motion.

c. **Operation.**

(1) To measure the elevation of the howitzer, place the proper reference surface of the quadrant on the leveling plates, parallel to the bore of the weapon, with the associated arrow pointing in the direction of fire. Set the level to indicate "ZERO" on the fine scale. Disengage the plunger from the notches in the frame, lift the arm and slowly lower it until the bubble is seen to pass through the central point. Allow the plunger to engage with the notches, then slide the level guide along the arm or turn the micrometer knob until the level bubble is accurately centered. Face the side of the quadrant which bears the arrow in use and read the coarse and fine scales. (In reading the M1 Quadrant, a zero indication is read as "0 mils" when the auxiliary indexes are matched, or as "10 mils" when they are not matched. Read red or black figures according to the engraved instructions below the micrometer.) The elevation of the howitzer in mils is equal to the sum of the coarse and fine scale readings. Remove the quadrant from the howitzer before firing.

(2) To measure depression angles, proceed as above, but with the arrow pointed in the reverse direction.

(3) To lay the howitzer to a given elevation, set the coarse and fine scales to the required angle and place the corresponding reference surface on the leveling plates. Elevate the howitzer, then depress it until the level bubble is centered. Remove the quadrant from the howitzer before firing.

d. **Test and Adjustment.** No adjustment of the quadrant by the using arms is permitted. The zero indication may be verified by setting the quadrant to zero elevation, elevating or depressing the howitzer to center the bubble, then turning the quadrant end for end. If the bubble is not centered, determine the elevation or depression angle necessary to center it; one-half of this angle is the error, and a corresponding correction should be applied to all subsequent indications in the 0- to 800-mil region.

e. **Care and Preservation.**

(1) Exercise particular care to prevent burring, denting, or nicking of the reference surfaces and of the notched portion of the frame.

(2) Do not leave the quadrant on the howitzer when firing.

(3) When not in use, keep the quadrant in the chest provided, with the shoes, forming the reference surfaces, lightly greased.

76. BORE SIGHT.

a. The bore sight (fig. 90) is used to indicate the direction of the axis of the bore of the howitzer, for orientation. Each bore sight is com-

155-MM HOWITZER M1 AND 155-MM HOWITZER CARRIAGE M1

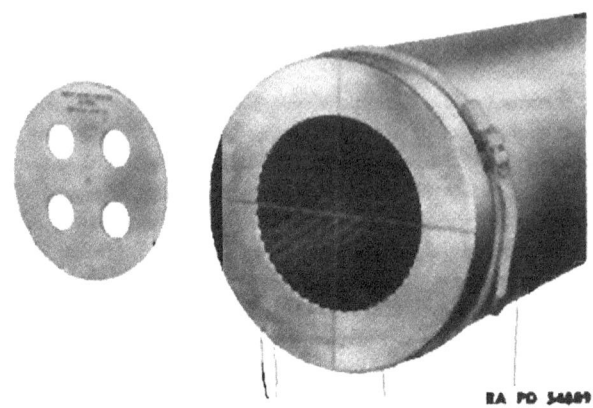

Figure 90 — Bore Sight

posed of a breech element and a muzzle element. The breech bore sight is a disk which fits accurately in the breech chamber of the howitzer. The model of the howitzer for which it is to be used is engraved on the disk. The muzzle bore sight includes a quantity of black linen cord to be stretched tightly across the muzzle, vertically and horizontally in the score marks thereon, and a web belt to be buckled around the muzzle to hold the cord in place.

b. With 2 elements in place, look through the central aperture in the breech bore sight. The direction of the axis is indicated by the intersection of the cords. See paragraph 78 on procedure for bore sighting.

c. Handle the breech bore sight carefully to prevent nicks and burs. Wind the cord and web belt into a compact bundle when it is not in use.

77. TESTING TARGET.

a. The testing target (fig. 91) is used during the bore-sighting operation (par. 8) for the alinement of sights and subcaliber equipment with the axis of the bore of the howitzer. The several aiming points are plainly designated. It is essential that the proper aiming points be selected for the materiel and equipment employed, and that the target be positioned in a vertical plane when in use. The normal distance from the howitzer at which the target should be located is about 50 yards.

78. PROCEDURE FOR BORE SIGHTING.

a. Level the carriage transversely (axis of trunnions) and longitudinally (axis of bore). Place the testing target about 50 yards away from and in a plane perpendicular to the bore of the howitzer. Place the bore sights in the howitzer and move the testing target until the center line

SIGHTING EQUIPMENT

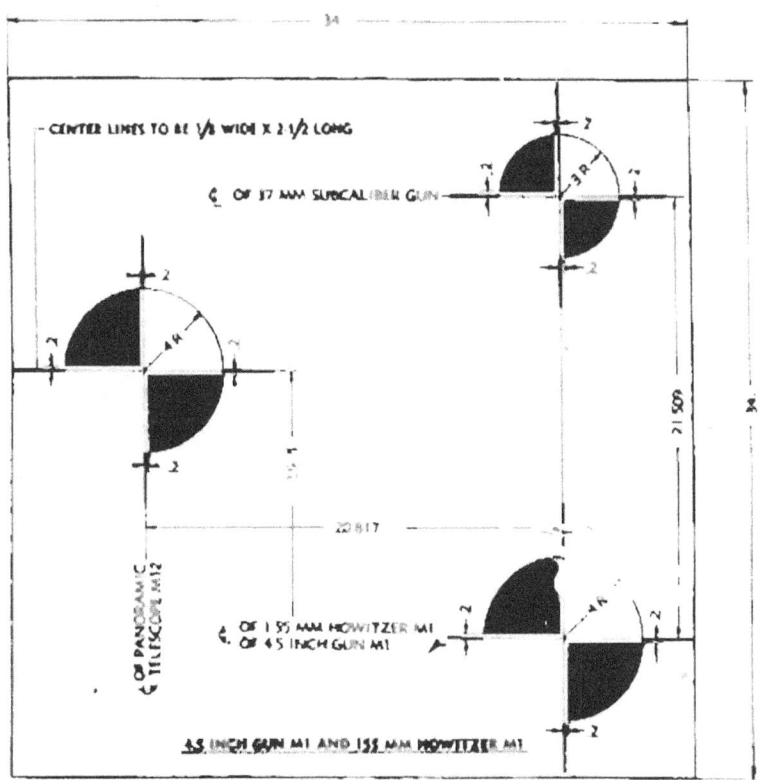

Figure 91—Testing Target

of the bore, as determined by the bore sights, pierces the lower circle of the testing target. The placement of the target in the vertical plane should be verified by a plumb line. Set the panoramic telescope azimuth, deflection, and elevation at "ZERO." The circle corresponding to the line of sight of the panoramic telescope should then appear exactly at the intersection of the crosslines on the reticle.

b. Failure of this circle and the crosslines to coincide vertically indicates that the elevating knob of the panoramic telescope is out of adjustment. Rotate the knob until the line of sight pierces the target on the horizontal center line. Loosen the locking screw at the center of the knob, slip the knob until the "ZERO" is in line with the index and tighten the locking screw. Do not lift the knob during this adjustment as such movement may cause disarrangement of the stop rings within the knob.

c. Failure of the circle and reticle to coincide laterally indicates that

TM 9-331

78

155-MM HOWITZER M1 AND 155-MM HOWITZER CARRIAGE M1

the azimuth micrometer is out of adjustment. Rotate the azimuth worm knob until the line of sight pierces the corresponding target on the vertical center line. Loosen the locking screw at the center of the micrometer index and slip the index until the arrow is in line with the "ZERO" of the deflection know (set to "ZERO" against the deflection index). Tighten the locking screw. If the azimuth scale and micrometer index do not both indicate "ZERO" simultaneously, or if the course and fine elevatin indexes do not indicate simulaneously, the matter should be brought to the attention of qualified ordnance personnel.

TM 9-331
79-80

Section II

FIRE-CONTROL EQUIPMENT

	Paragraph
General	79
Aiming circle M1	80
Compasses	81
One-meter base range finder M1916	82
Fuze setter M14	83
Graphical firing table M14	84
B.C. telescope M1915 or M1915A1	85

79. GENERAL

a. The fire-control equipment used with this weapon includes the following items:

(1) The Aiming Circle M1 (with Instrument Light M2) used for measuring angles in azimuth and site and for general topographical work.

(2) The compass used for measuring angles of site, chronometer angles and magnetic azimuths. The Prismatic Compass M1918 (Sperry) will be used until the Compass M2 becomes available.

(3) The 1-meter Base Range Finder M1916 used for measuring distances by triangulation. Azimuths and angles of site may also be obtained.

(4) The Fuze Setter M14 used for setting the mechanical time fuzes of projectiles for this materiel.

(5) The Graphical Firing Table M13 (short range) and M20 (long range), used to simplify and speed up the conduct of fire, and to help reduce the probability of error.

(6) The Battery Commander's Telescope M1915 or M1915A1 (with Instrument Light M1) used for observation and for measurement of azimuth and angle of site.

80. AIMING CIRCLE M1.

a. Description. This instrument is used for measuring angles in azimuth and site, and for general topographical work. It includes a 4-power telescope with a laterally and vertically graduated reticle, 2 levels, a declinator, elevating, orienting, and azimuth mechanisms, and azimuth scales and micrometers. Azimuth indications are in miles, numbered to correspond to the scale indications of the other instruments commonly used with the aiming circle. No scale other than that on the reticle is provided for vertical angles. The instrument is furnished complete with tripod and carrying case (figs. 92, 93 and 94).

TM 9-331
80

155-MM HOWITZER M1 AND 155-MM HOWITZER CARRIAGE M1

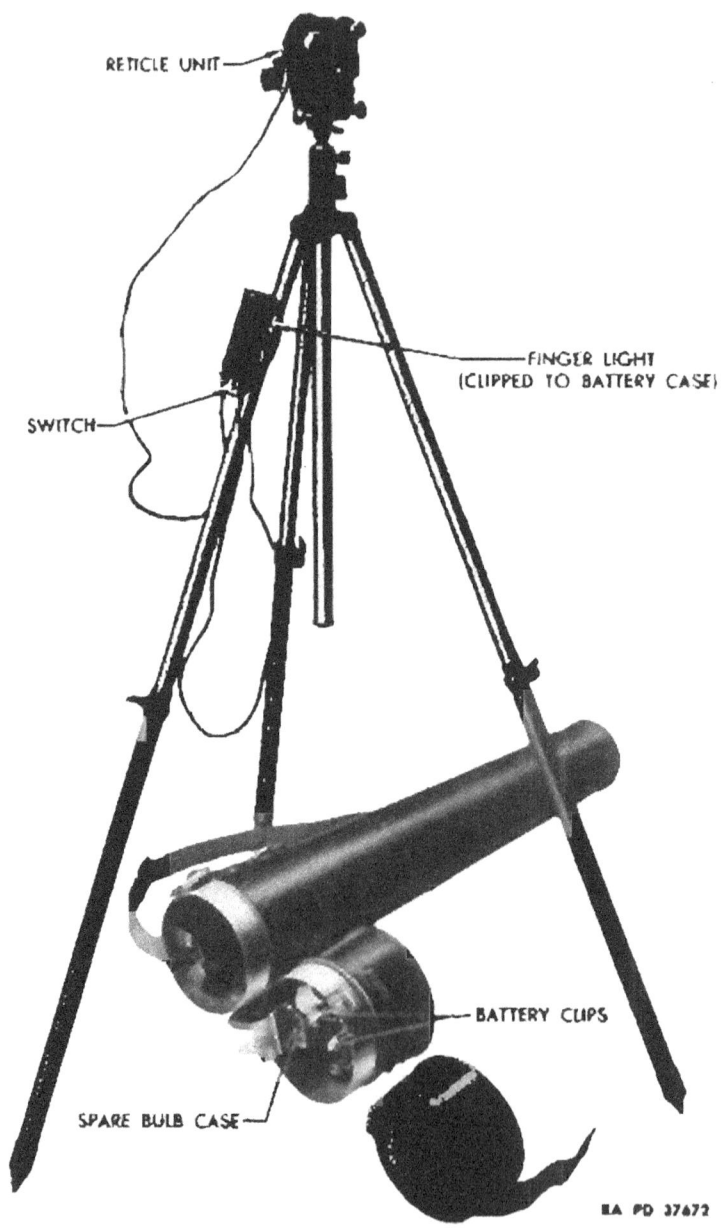

Figure 92 — Aiming Circle M1 — With Instrument Light M2 and Carrying Case

FIRE-CONTROL EQUIPMENT

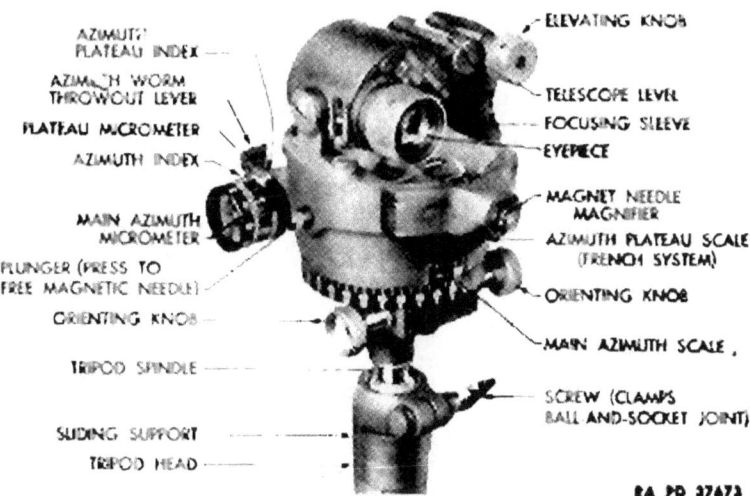

Figure 93 — Aiming Circle M1 — Left Rear View

b. Description of Instrument Light. All Aiming Circles M1 are being equipped with the Instrument Light M2, which includes a battery case, connected by flexible cords to a reticle unit, and a finger light. The battery case, containing one flashlight cell, is clamped to a tripod leg and has a switch controlling both its lamps. The reticle unit snaps in place in a dovetailed slot over the reticle illuminating window. The finger light has a soft rubber housing, and is held in a spring clip on the battery case, when not in use. The aiming circle carrying case is being modified to permit storage of the flashlight cell and spare cell and spare bulbs separately from the battery case (fig. 92).

c. Operation.

(1) To set up the instrument, clamp the tripod legs at the desired length, and embed them firmly in the ground. Clamp the sliding support of the tripod at the desired height. Level the instrument, using the circular level and the ball-and-socket joint. Focus the telescope as required, using the sleeve on the eyepiece.

(2) To orient the instrument, either a datum point of known azimuth, or magnetic bearings may be used.

(a) To orient on a datum point of known azimuth, set the main azimuth scale (100-mil steps) and micrometer (1-mil steps) to the azimuth of the datum point, and turn one of the orienting knobs until the datum point appears on the vertical crossline of the reticle. The instrument may also be relocated on the tripod spindle, using the orienting clamping screw for large angular changes. The telescope may be elevated or depressed, as required, to bring the point in the field of view.

155-MM HOWITZER M1 AND 155-MM HOWITZER CARRIAGE M1

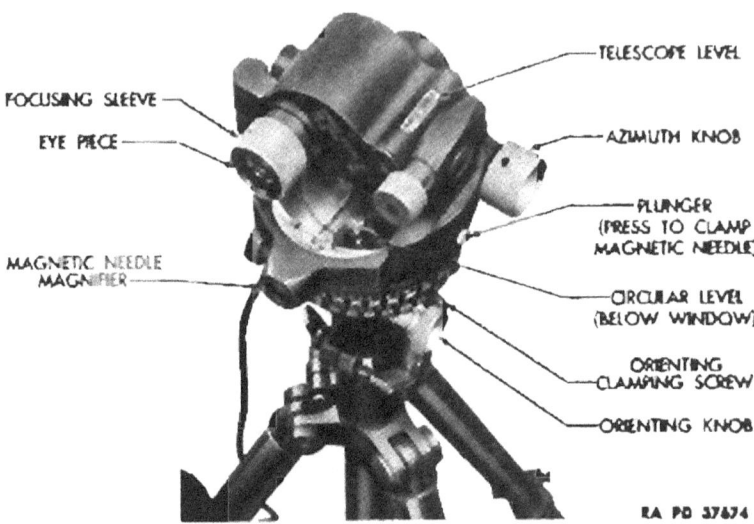

Figure 94 — Aiming Circle M1 — Top Right Rear View

(b) To orient on magnetic north, set the main azimuth scale and micrometer to indicate "ZERO." Press the plunger releasing the magnetic needle, and turn one of the orienting knobs until the north-seeking (knife edge) end of the magnetic needle appears approximately opposite the "N" index at the front of the instrument, then refine the setting so that the south-seeking (rectangular) end of the needle is centered in the reticle, viewed through the magnifier. The instrument may also be relocated on the tripod spindle, using the orienting clamping screw for large angular changes. The aiming circle will then indicate magnetic azimuths.

(c) To orient on grid north, proceed as for magnetic north, but set the azimuth to the magnetic declination of the locality (subtracting west declinations from 6,400 mils) instead of to zero. The instrument will then indicate grid azimuths.

(d) When orientation by magnetic bearings has been completed, press the red plunger to clamp the magnetic needle.

(3) To read angle of site, rotate the elevating knob so that the bubble of the telescope level is centered. The angle of site of an object is then indicated by its position on the graduations at 5-mil intervals along the vertical crossline of the reticle. Angles of site, thus measured, are limited to 85 mils and no other indicating means is provided.

(4) To read azimuth, bring the object on the vertical crossline of the reticle, using the azimuth knob; the throwout lever may be depressed for making large azimuth changes rapidly. The azimuth indications of this

FIRE-CONTROL EQUIPMENT

instrument may be read either directly in mils, or in terms of the indications on the panoramic telescope as follows:

(a) Azimuths from 0 to 6,400 mils are read directly on the azimuth scale, using the main (upper) graduations for values from 3,200 mils up. Indications on this scale are at 100-mil intervals, and are supplemented by those on the azimuth micrometer, which is graduated at 1-mil intervals.

(b) Angular indications corresponding to those of the Panoramic Telescope M25 (0-3,200, 0-3,200-mil scales) are similarly read, using the auxiliary (lower) graduations from azimuths for over 3,200 mils.

(c) Small angles may be measured along the horizontal crossline of the reticle, which is graduated at 5-mil intervals.

(d) The azimuth plateau scale and micrometer are for use with the sighting equipment on certain 75-mm gun carriages.

(5) To prepare the instrument for traveling, place it in the carrying case provided. The instrument need not be removed from the tripod.

d. Tests and Adjustments.

(1) The azimuth and plateau micrometers should read "0" and "100," respectively, when the azimuth scale indicates "ZERO." Three screws in the end of the azimuth micrometer may be temporarily loosened for this adjustment.

(2) The telescope level should indicate the line of sight, determined by the center of the reticle, to be horizontal. This may be verified by sighting on a distant point at the same level as the telescope, the error, if any, being read on the reticle. No corrective adjustment by the using arm is permitted. A celluloid strip is provided on the front of the instrument for recording corrections.

(3) To check the accuracy of the declinator, it is necessary to set up the instrument in a position not subject to local magnetic attraction, and to sight on one or (preferably) more points of known azimuth. The average error should be noted and the necessary correction recorded on the celluloid strip. No adjustment by the using arm is permitted.

e. Care and Preservation.

(1) Refer to paragraph 71 for general instructions pertaining to the care and preservation of instruments.

(2) Exposed moving parts should be oiled occasionally. Interior parts are not to be lubricated by the using arms. Keep excessive lubricant that seeps from the mechanisms wiped off to prevent accumulation of dust and grit.

(3) When storing aiming circles equipped with instrument lights, remove the flashlight cell from the battery case and place it in the compartment of the aiming circle carrying case.

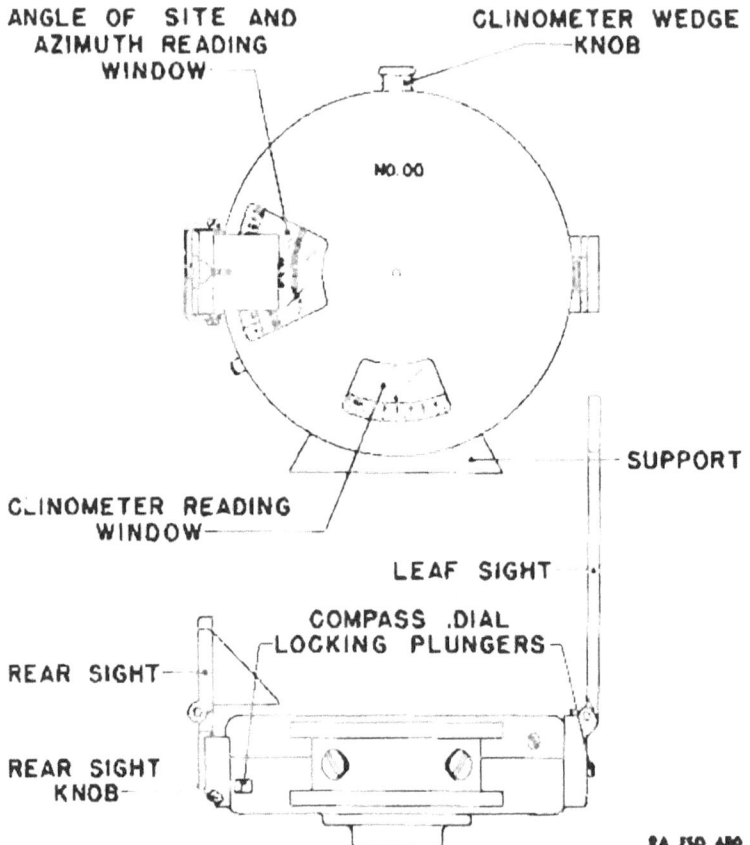

Figure 95 — Prismatic Compass M1918 (Sperry)

81. COMPASSES.

a. Prismatic Compass M1918 (Sperry). This instrument (fig. 95) is used for measuring angles of site, clinometer angles, and magnetic azimuths. The Prismatic Compass M1918 (Sperry) will be used until the Compass M2 which is the present standard becomes available (subparagraph *b* below).

(1) DESCRIPTION. This instrument is furnished complete with a carrying case but without a tripod. The instrument includes a compass dial (green) carrying a magnetic needle and azimuth scales, a weighted clinometer dial (white), and a sighting system whereby angular indications may be read while observing the object.

(2) OPERATION.

(a) To measure angles of site, raise leaf sight and rear sight. Pull out

TM 9-331
81

FIRE-CONTROL EQUIPMENT

clinometer wedge knob to permit free rotation of clinometer dial. Focus rear sight on clinometer (white) dial, sliding sight as required and clamping it in position with rear sight knob. Hold instrument with dials in a vertical plane, look through niche in rear sight, and elevate or depress instrument until object observed is in line with horizontal central vane of leaf sight. The angle of site, reflected in rear sight prism, will also be visible in the center of the field of view. The angle of site scale (outer scale on clinometer dial) is graduated at 5-mil intervals and numbered at 100-mil intervals. The 50-mil points are also marked. A 300-mil indication corresponds to a level line of sight, as on the corresponding scales of range quadrants. The clinometer wedge knob may be partially depressed to damp out oscillations. It must not be depressed when taking the reading.

(b) To measure azimuths, first operate the instrument in angle of site until the compass (green) dial is exposed at rear sight by cutaway portion of clinometer (white) dial. Depress clinometer wedge knob. Raise leaf sight and rear sight. Focus rear sight on compass (green) dial, sliding sight as required and clamping it in position with rear sight knob. Hold instrument in hand or support it on a convenient nonmagnetic body. Look through niche in rear sight and rotate instrument in azimuth until the object observed is in line with the vertical central vane of the leaf sight. The magnetic azimuth, reflected in rear sight prism, will also be visible in the center of the field of view. The compass dial is graduated at 10-mil intervals and numbered at 100-mil intervals. Additional numbering is provided in the 3,200 to 6,400-mil half of the scale to correspond to the numbering on azimuth scales of panoramic telescopes which are graduated 0 to 3,200 mils in this range. To damp out oscillations of compass dial, gently depress one of the locking plungers. Plungers must not be in depressed position when taking the azimuth reading.

(c) To use the instrument of a clinometer, pull out clinometer wedge knob and stand instrument, prism to the rear, on its support, on a straight portion of the piece which is parallel to the bore. The reading of clinometer scale, read opposite on etched line of clinometer reading window, is the elevation of the piece. The clinometer scale is graduated at 10-mil intervals and numbered at 100-mil intervals. A 300-mil reading indicates bore of the piece to be level. Sights should not be raised when using instrument only as a clinometer. The clinometer wedge knob may be partially depressed to damp out oscillations. It must not be depressed when taking the reading.

(d) To prepare instrument for travel, push in clinometer wedge knob (clamping clinometer dial) and turn leaf sight down (clamping the compass dial). Lower and fold back rear sight, securing it in place with the catch. Place instrument in case provided.

(3) TEST AND ADJUSTMENT. Accuracy of azimuth and angle of site

TM 9-331
81

155-MM HOWITZER M1 AND 155-MM HOWITZER CARRIAGE M1

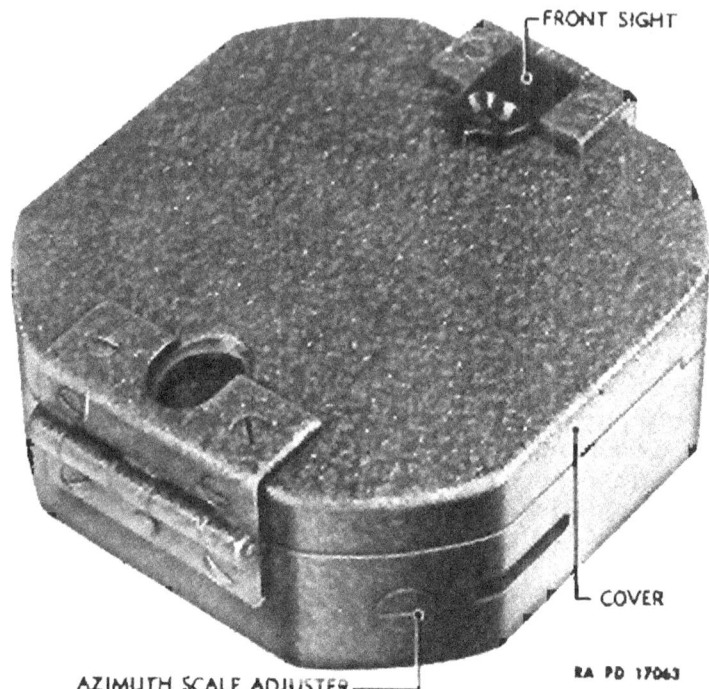

Figure 96 — Compass M2 — Cover Closed

indications may be checked by sighting on datum points of known azimuth and elevation. When placed on a flat level surface, the clinometer should indicate 300. No adjustment by the using arms is permitted.

(4) CARE AND PRESERVATION.

(a) See paragraph 71 for general instructions pertaining to care and preservation of instruments.

(b) When not in use keep leaf sight down, clamping compass dial, and clinometer wedge knob depressed, clamping clinometer dial, thus preventing injury to and excessive wear of their respective pivots.

(c) Observe particular care to prevent bending of the leaf sight parts.

(d) No lubrication of the instrument is required.

b. Compass M2.

(1) GENERAL. The Compass M2 (figs. 96 and 97) is a multiple-purpose instrument used for measuring angles of site, clinometer angles, and magnetic azimuths. It has been adopted as standard to replace the Prismatic Compass M1918 (Sperry), which has been reclassified as limited standard.

(2) DESCRIPTION. The compass weighs about 8 ounces, and measures

142

FIRE-CONTROL EQUIPMENT

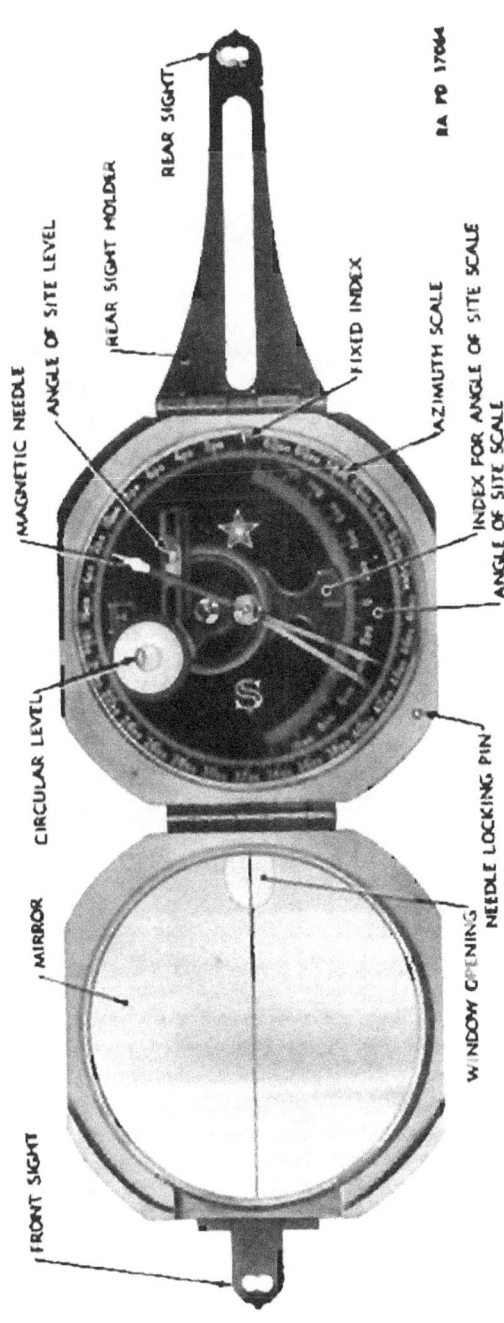

Figure 97 — Compass M2 — Cover Open

TM 9-331
81

155-MM HOWITZER M1 AND 155-MM HOWITZER CARRIAGE M1

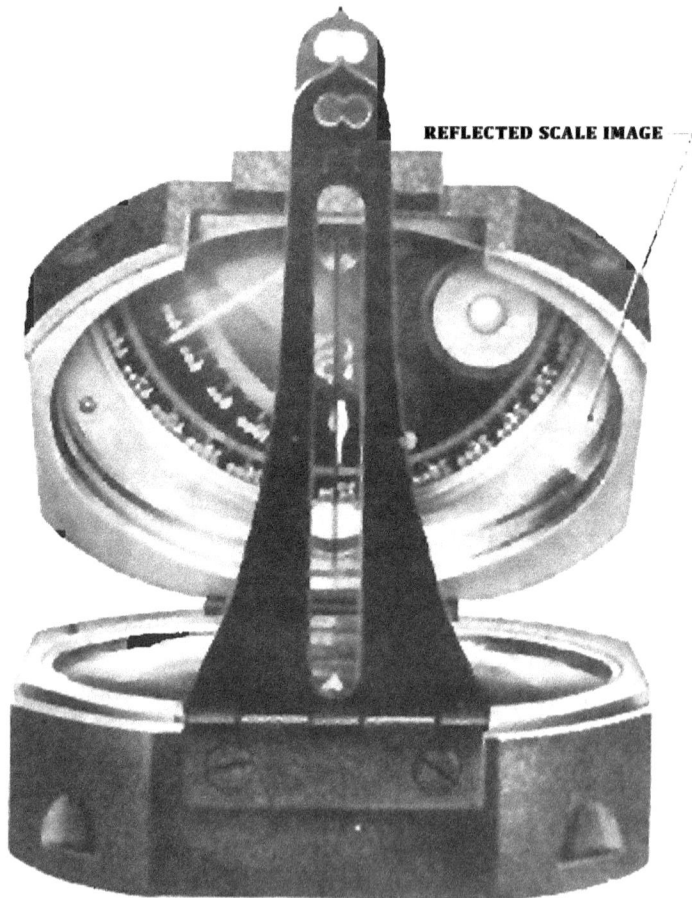

Figure 98 — Compass M2 — Observer's View, Azimuth

over-all about 2½ x 2½ x 1½ inches when closed. It consists of a compass and angle of site mechanism contained in a body with a hinged cover. When the cover is closed, it lifts the magnetic needle from its pivot and clamps it securely for traveling. The north-seeking end of the magnetic needle is painted white. The compass scale can be oriented on grid (Y) north for any locality. Front and rear sights together with a mirror in the cover, permit measurements to be made while observing the object.

(3) ACCESSORIES. The compass is furnished complete with Carrying Case M19. The case is velveteen-lined and has a leather loop on the back for carrying on the user's belt.

FIRE-CONTROL EQUIPMENT

(4) OPERATION. Compass operating positions are shown in figures 98 and 99. The compass should be held as rigidly as possible to obtain the most accurate readings. The use of a sitting or prone position, a rest for the hands or elbows, or a solid support for the compass, will help to eliminate unintentional movement of the instrument. When measuring angles in azimuth, the compass should be used away from steel or iron masses which would distort the local magnetic field. The time of oscillation of the magnetic needle can be shortened by partially depressing the needle locking pin, or the average position of the needle in its swing may be used. Practice in the operation of the compass greatly increases the proficiency and accuracy attained by the operator.

(5) TO MEASURE ANGLES IN AZIMUTH BY READING REFLECTED IMAGE OF AZIMUTH SCALE.

(a) Hold the opened compass in both hands at eye level, with arms braced against body and the rear sight nearest the eyes. Place the cover at an angle of approximately 45 degrees to the face of the compass (fig. 98) so that the reflected scale can be readily viewed. Level the instrument by means of the reflected circular level, sight on the desired object, and read the azimuth in the reflected image of the scale. The azimuth reading is indicated on the azimuth scale by the south-seeking (black) end of the compass needle. When sighting, hold hands rigid and turn body. The instrument can be sighted by any of the methods below. More accurate readings result from the use of a longer sight base.

(b) Raise the rear sight holder approximately perpendicular to the face of the compass. Sight on the object through the opening in the rear sight holder and through the window in the cover (fig. 98). Keep the compass level and raise or lower the eye along the opening in the rear sight holder until the black center line of the window bisects the object and the opening in the rear sight.

(c) Fold the rear sight holder out parallel with the face of the compass, with the rear sight perpendicular to its holder. Sight through or over the rear sight and view the object through the window in the cover. If the object sighted is at a lower elevation than the compass, raise the rear sight holder as needed. The compass is correctly sighted when the compass is level and the operator sees the black center line of the window bisecting the rear sight and the object sighted.

(d) Raise the front sight and the extended rear sight assembly perpendicular to the face of the compass. Sight over the tips of the rear and front sights. If the object is above the line of sighting, fold the rear sight toward the eye as needed. The instrument is correctly alined when, with the level centered, the operator sees the tips of the sights and the center of the object sighted in coincidence.

TM 9-331
155-MM HOWITZER M1 AND 155-MM HOWITZER CARRIAGE M1

Figure 99 — Compass M2 — Observer's View, Site

TM 9-331
81

FIRE-CONTROL EQUIPMENT

(6) TO MEASURE ANGLES IN AZIMUTH BY READING AZIMUTH SCALE DIRECTLY.

(a) Hold the opened compass in both hands (at about waist level), braced against the body, with the rear sight away from the body. Open the cover until the mirror affords a clear image of the object sighted. Extend the rear sight and raise the rear sight assembly until it is approximately perpendicular to the face of the compass. Level the instrument by means of the circular level. Holding arms rigid with the instrument level, turn body until the centerline on the mirror bisects the opening in the rear sight holder and the mirror image of the object sighted.

(b) Then read the azimuth indicated on the azimuth scale by the north-seeking (white) end of the compass needle.

(7) TO MEASURE ANGLES OF SITE.

(a) Hold the opened compass in a vertical plane as in figure 99, with the rear sight toward the body and the angle of site level lever to the right. Open the cover to an angle of approximately 45 degrees to the face of the compass. Fold the rear sight holder out parallel to the face of the compass with the rear sight perpendicular to the holder.

(b) Look through the rear sight and raise or lower the instrument until the center line of the window bisects the opening in the rear sight and the object sighted.

(c) Then level the tubular level reflected in the mirror, by means of the lever. Open the cover and read the angle of site opposite the index.

(d) Care must be exercised to maintain the compass in a vertical plane to obtain accurate readings.

(8) TO MEASURE CLINOMETER ANGLES. Open the cover and rear sight holder parallel with the face of the compass. Place the edge of the opened compass on the leveling plates of the piece, center the bubble of the tubular level, and read the angle of site.

(9) ORIENTATION ON GRID (Y) NORTH.

(a) The standard reference direction for compass readings is grid (Y) north, corresponding to the grid indications on standard maps. However, due to regional differences in magnetic direction and local disturbances in the magnetic field, the magnetic needle in the compass may point several degrees to either side of the reference direction. The difference between the magnetic direction and the standard reference direction is the declination constant.

(b) Allowance for declination constant can be made in this compass by orienting (shifting) the azimuth scale, using the azimuth scale adjuster. The slotted head of the adjuster can be turned with an ordinary screwdriver.

(c) To determine the declination constant, open the compass and set zero of the azimuth scale against the fixed index in the body by means of the azimuth scale adjuster. Take compass readings on several points of known azimuth. Compute the difference between the compass reading (mean of 3 readings) of each of the points and the known grid (Y) azimuth. The mean of these differences is the declination constant of the instrument for the particular locality.

(d) If the azimuth readings are greater than the grid azimuths, subtract the declination constant from the azimuth readings or rotate the azimuth scale the amount of the declination constant in a counterclockwise direction by means of the azimuth scale adjuster. If the azimuth readings are less than the grid azimuths, add the declination constant to the azimuth readings or rotate the azimuth scale in a clockwise direction. This orients the compass on grid (Y) north.

(e) An alternate method of finding the declination constant is to use an isogonic chart. This method is less accurate as it does not consider local disturbances in the magnetic field.

(f) If the compass is to be used in another locality 6 or more miles distant, declination constant should be determined for new locality.

(10) ADJUSTMENT. Adjustment for dip of the magnetic needle and errors in the circular and tubular levels, may not be made by the using arm personnel. However, errors in the tubular level may be determined by comparison against a level or gunner's quadrant of known accuracy. If the error remains constant, it can be compensated for in measuring angles of site or when using the instrument as a clinometer.

(11) CARE AND PRESERVATION.

(a) The compass should be handled carefully to avoid unnecessary shocks. It should be closed and kept in the carrying case when not in use. After use in wet weather, wipe the compass dry before placing it in the carrying case.

(b) When the instrument is moved from one position to another, or is not in use, close the cover, locking the needle off its pivot. This prevents injury to the needle pivot.

(c) Particular care should be exercised to prevent bending the sights or the cover hinge. Lay the rear sight flat before closing the cover.

(d) Moisture due to condensation may collect in the instrument when the temperature of the parts is lower than that of the surrounding air. This moisture, if not excessive, can be removed by placing the instrument in a warm place.

(e) No lubrication is required.

82. ONE-METER BASE RANGE FINDER M1916.

a. General. This instrument (figs. 100 and 101) is used primarily for measuring distance by triangulation. Indications of azimuth and angle of site are also provided.

b. Description. The instrument includes an internal 1-meter base

FIRE-CONTROL EQUIPMENT

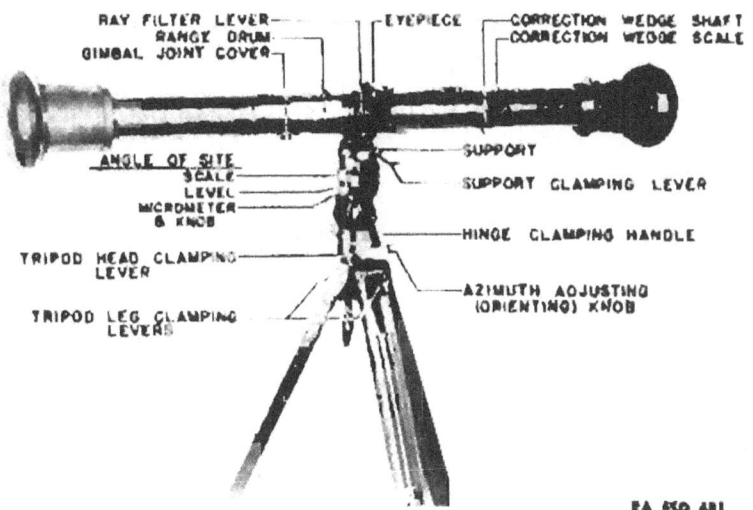

Figure 100—Range Finder M1916—1-Meter Base—Rear View

line, a 15-power optical system with 2 objectives and a common eyepiece of the coincidence type, and a scale on which the distance is indicated. It is furnished complete with mount and tripod. Necessary carrying cases and adjusting equipment are provided as accessories. The mount positions the line of sight of the range finder in elevation and azimuth and provides a hinge joint for placing the base line axis of the instrument in either a vertical or horizontal position. Angle of site and azimuth scales and micrometers are provided on the mount.

c. Operation.

(1) To set up the instrument, clamp tripod legs securely at desired length, embed them firmly in the ground, and tighten leg clamping levers. It is necessary that azimuth scale is in a substantially horizontal plane. Place range finder on mount and latch it in position. Position longitudinal axis horizontally and clamp with hinge clamping handle. Procedure for a vertical base line is described in subparagraph (5).

(2) To prepare optical system for use, rotate end box sleeves, uncovering both windows. Set ray filter lever to proper position. No filter at all may be used, or the amber filter (for exceptionally bright daylight or reflection of sun over water), or the smoked filter (for observation near the sun or into direct rays of a searchlight) may be employed. Focus eyepiece by rotating diopter scale to produce a clear image. If the operator knows the value of his own eye the setting may be made directly on the scale.

TM 9-331
82

155-MM HOWITZER M1 AND 155-MM HOWITZER CARRIAGE M1

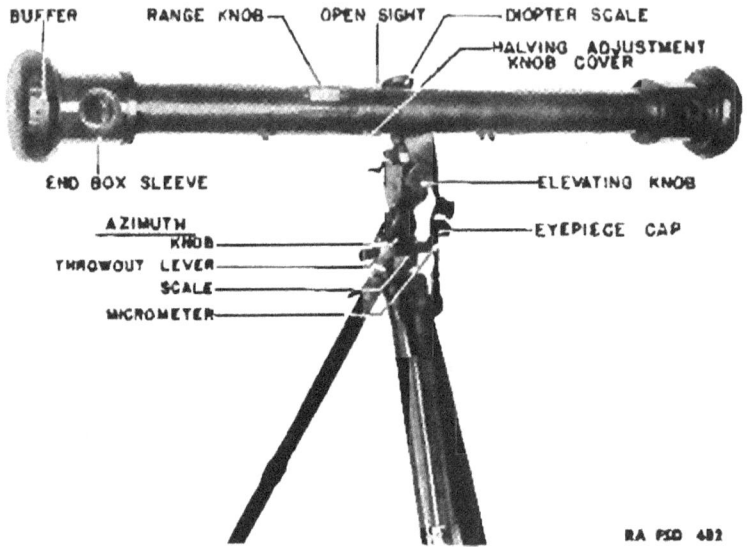

Figure 101 — Range Finder M1916 — 1-Meter Base — Front View

(3) To orient the instrument, select a datum point of known azimuth. Set this value on azimuth scale and micrometer. Loosen tripod head clamping lever and swing instrument until datum point appears near vertical centerline of field of view, indicated by a short line in lower field of view. Clamp lever and refine setting with azimuth adjusting (orienting) knob so that point appears exactly on vertical center line.

(4) To measure the range of an object, select a clearly defined part perpendicular, if possible, to the halving line. Move the instrument in azimuth and elevation as required to bring the part at center of field of view when in coincidence. On moving targets it is advisable to start with the target at edge of field of view so that it may be brought into coincidence as it crosses the field. An open sight is provided for picking up the target. For large angular displacements in azimuth, depress throwout lever and turn instrument as required. When first observed, images will ordinarily not be in coincidence (fig. 102 (1)). Turn range knob until image of point selected appears in coincidence (fig. 102 (2)). Read range in yards on range drum opposite sliding range pointer.

(5) To measure range of horizontal objects such as roads, trenches, crests of ridges, etc., which have no prominent vertical parts, turn instrument with longitudinal axis vertical, temporarily loosening hinge clamping handle for the purpose. Images when first observed will ordinarily not be in coincidence (fig. 103 (1)). Turn the range knob until the

FIRE-CONTROL EQUIPMENT

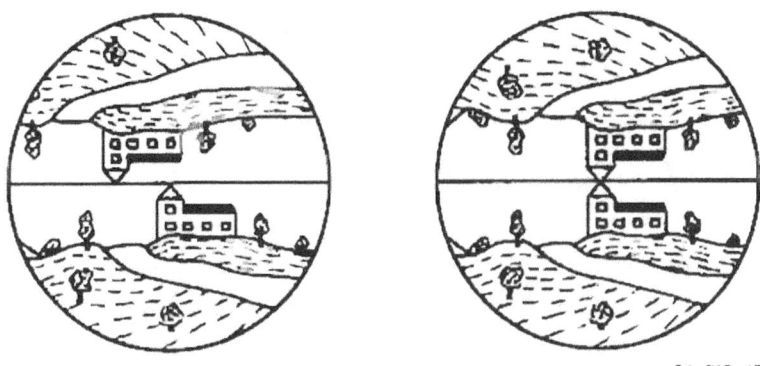

Figure 102 — Range Finder Field of View — Horizontal Base

image of horizontal line appears to continue across the halving line as at A in figure 103 (2). Temporarily lower the support clamping lever for any necessary motion in elevation (within limits of ± 10 deg).

(6) To read angle of site, center level bubble, using angle of site knob. The angle of site indication may then be read on associated scale (100-mil steps) and micrometer (1-mil steps). An indication of 300 mils corresponds to a horizontal line of sight. Angle of site can be read only when using the instrument with longitudinal axis horizontal.

(7) To read azimuth, the azimuth scale (100-mil steps) and micrometer (1-mil steps) furnish the necessary indications. It is essential that the plane of the azimuth scale be substantially level and that the object

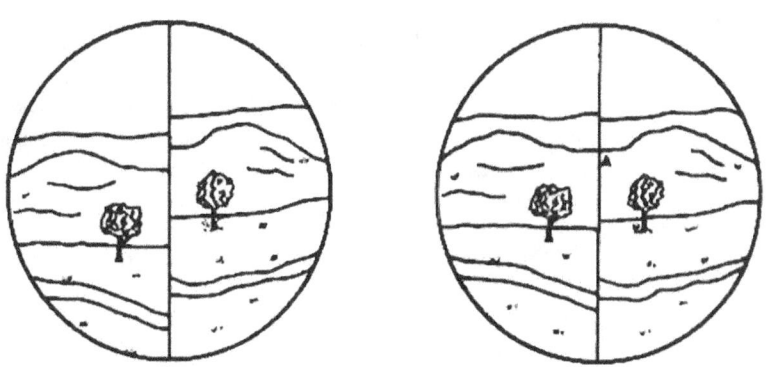

Figure 103 — Range Finder Field of View — Vertical Base

TM 9-331
82

155-MM HOWITZER M1 AND 155-MM HOWITZER CARRIAGE M1

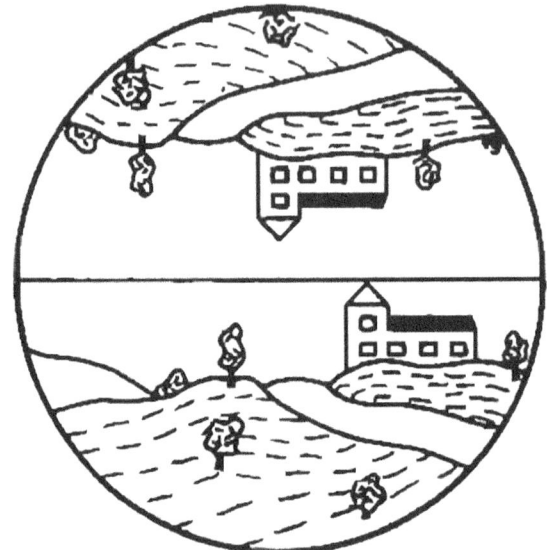

Figure 104 — Range Finder Field of View — Incorrect Halving

be at the center of field of view for correct angular indications. Azimuth may be measured with longitudinal axis either vertical or horizontal, but the instrument must be oriented separately for each position. Azimuths from 3,200 to 6,400 mils have an additional auxiliary scale reading from 0 to 3,200 mils for use with panoramic telescopes similarly graduated.

(8) To prepare instrument for traveling, cover the eyepiece, close the end box sleeves and the cover over range drum. Remove range finder from mount and place in its case. Place mount and tripod in case, with elevating knob toward inside of case. Do not remove mount from tripod. Remove sight from adjusting lath. Place lath in internal pocket of tripod carrying case and sight in lid pocket. The latter pocket also contains the correction wedge key and a camel's-hair brush.

d. Test and Adjustment.

(1) HALVING LINE. Incorrect adjustment of the halving line is indicated by failure of the corresponding points on inverted and erect images to fall on the halving line (fig. 104). To correct the halving, slide back cover exposing the halving adjustment knob and rotate knob until the corresponding point on each image touches the halving line ((1) and (2), fig. 102). A sharply defined point at large 400 yards away must be used for this adjustment. Return cover to its original position when adjustment is completed.

FIRE-CONTROL EQUIPMENT

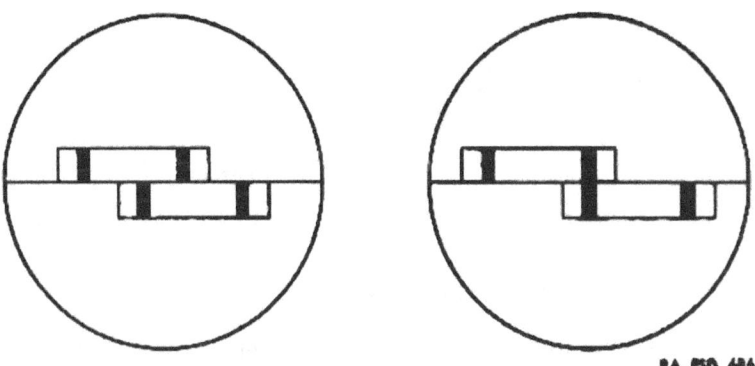

Figure 105 — Range Adjustment — Infinity Method

(2) RANGE INDICATION.

(a) To test the instrument using a finite range, select a sharply defined object at a distance of 400 yards or more, the range of which is accurately known, and bring the object into coincidence in the center of field of view ((2), fig. 102). If range adjustment is correct, the known range should be indicated.

(b) To test using moon or other celestial body (never the sun), proceed as for an object at finite range. Infinite range (00) should be indicated.

(c) To test the instrument by the infinity method, prepare adjusting lath by inserting the sight (carried in pocket flap of carrying case). Place adjusting lath in a horizontal position 200 yards or more from the instrument. Use the sight on the lath to insure perpendicularity to line of sight. Set range drum to indicate infinite range (00). If images appear alined as in (2), figure 105, the adjustment is correct. Misalinement, such as is shown in (1), figure 105, indicates necessity for adjustment.

(d) To adjust instrument in range, set range at the known range or at infinity, depending on the method of test employed, and bring images into correct relation, using the correction wedge key to turn correction wedge shaft. Note indication on correction wedge scale, repeat several times, and set the scale to the average of the readings.

(e) It is essential that the adjusting lath, when used, is the one belonging with the instrument. The same serial number is provided on both.

(3) AZIMUTH INDICATION. If azimuth scale and micrometer fail to indicate zero simultaneously, the latter may be slipped around as required, temporarily loosening clamping screw in the end.

(4) ANGLE OF SITE INDICATION. Sight on a point at least 400 yards distant at the same level as the range finder. The angle of site indication should be normal (300 mils). Correction for small errors may be applied

155-MM HOWITZER M1 AND 155-MM HOWITZER CARRIAGE M1

Figure 106—Fuze Setter M14

by slipping angle of site micrometer through required angle, temporarily loosening clamping screw in the end.

e. **Care and Preservation.**

(1) See paragraph 71 for general instructions pertaining to care and preservation of instruments.

(2) The gimbal joint cover is not to be removed by using arms.

(3) Keep cover over the halving adjustment knob closed except when making an adjustment.

(4) The range finder should not be pointed directly at the sun. This instrument contains a cemented prism which will be injured by such practice.

(5) Avoid striking or bumping the instrument at the ends when mounted, as the parts at the center will thereby be subjected to excessive stress.

(6) Exposed moving parts of mount should be oiled occasionally. Interior parts of mount and range finder are not to be lubricated by using arm. Keep excess lubricant that seeps from the mechanism wiped off to prevent accumulation of dust and grit.

83. FUZE SETTER M14.

a. The Fuze Setter M14 (fig. 106) is a flat steel wrench, with projecting pins, built to fit the contour of the fuze and engage the notches of the setting ring. It is used for manually setting the time on mechanical time fuzes or projectiles for this materiel.

84. GRAPHICAL FIRING TABLE M14.

a. Graphical firing tables (fig. 107) are used to simplify and speed up the conduct of fire, and to help reduce the probability of error. The graphical firing table consists of a graduated stock and slide to form a Mannheim-type slide rule. The range scale on the stock of the rule is plotted logarithmically. All of the scales are so plotted as to conform to this range scale. Graphical firing tables are made for each field artillery weapon, used for indirect fire, and are designated by model.

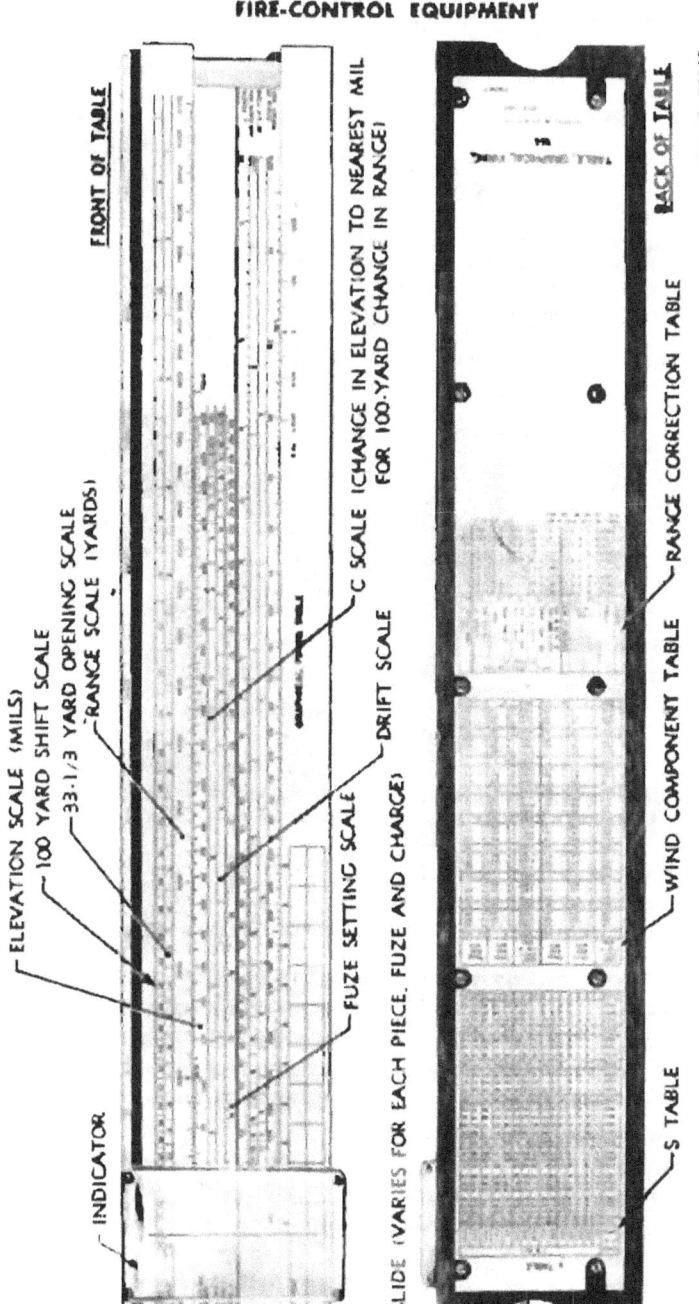

Figure 107 — Graphical Firing Table

TM 9-331
84-85

155-MM HOWITZER M1 AND 155-MM HOWITZER CARRIAGE M1

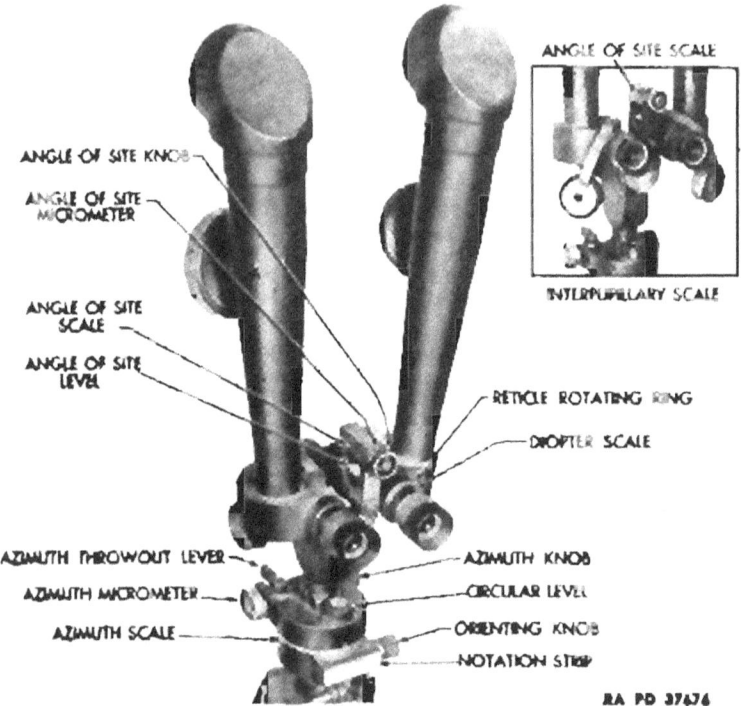

Figure 108—Battery Commander's Telescope M1915A1—Rear View, With Telescopes Positioned Vertically

b. **Graphical Firing Tables M13 and M20.** These are standard for use with the 155-mm Howitzer M1. The M13 is for short range and the M20 is for long range. Other information is not available at this time.

85. B.C. TELESCOPE M1915 OR M1915A1.

a. **General.** The Battery Commander's Telescope M1915 or M1915A1 (figs. 108 and 109) is a 10-power binocular instrument used for observation and for measurement of azimuth and angle of site. It is furnished complete with mount and tripod, and the necessary carrying cases, storage chest, and cleaning brushes.

b. **Description.**

(1) The telescopes are arranged so that they may be positioned vertically (fig. 108) or swung down horizontally so as to provide an accentuated stereoscopic effect (fig. 109).

(2) The Battery Commander's Telescope M1915A1 is equipped for reticle illumination and is designed to receive the Instrument Light M1. Illumination for the instrument may be supplied by flashlight until such

TM 9-331
83

FIRE-CONTROL EQUIPMENT

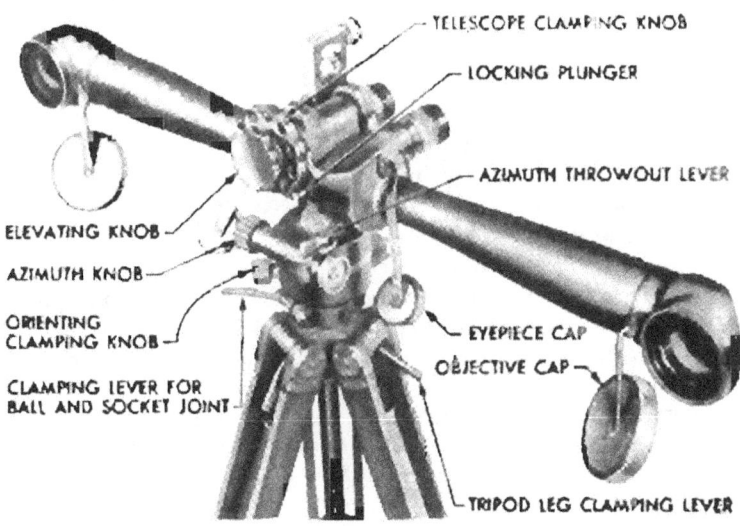

Figure 109—Battery Commander's Telescope M1915A1—Front View, with Telescopes Swung Down Horizontally to Provide an Accentuated Steroscopic Effect

time as the instrument light becomes available. Unmodified instruments, without means for reticle illumination, are designated M1915.

c. **Operation.**

(1) To set up the instrument, remove the tripod and mount from the tripod carrying case, clamp the tripod legs at the desired length, embed them firmly in the ground, and tighten the leg clamping levers. Remove the telescope from its carrying case. Place it on the vertical spindle extending from the mount, depressing the locking plunger, turning the telescope until the mating surfaces of telescope and mount engage properly, and then releasing the plunger. Level the mount, using the circular level and the ball-and-socket joint at the bottom of the mount. Clamp the mount with the lever when the level bubble is centered.

(2) Prepare the telescope, remove the caps from the eyepieces and objectives. If required, place the sunshades over the objectives and the amber filters over the eyelenses. Sunshades and filters are carried in compartments of the telescope case. Release the telescope clamping knob and turn the telescopes to the vertical or horizontal position as required. At the same time, set the proper interpupillary distance in millimeters on the associated scale, and clamp the scale in place. If the interpupillary distance for the observer is not known, it may be found by observing the sky and moving the eyepieces apart or together until the field of view

155-MM HOWITZER M1 AND 155-MM HOWITZER CARRIAGE M1

change from two overlapping circles to one sharply defined circle. Focus each eyepiece independently, looking through the telescope with both eyes open, at an object several hundred yards away, covering the front of one telescope and turning the diopter scale until the object appears sharply defined. Repeat this adjustment for the other eye. A diopter scale is provided for each eye, and if the observer remembers the values for his own eyes, the setting may be made directly on the scales. Turn the reticle rotating ring until the reticle appears erect.

(3) To orient the instrument, select a datum point of known azimuth and set this value on the azimuth scale (100-mil steps) and micrometer (1-mil steps). The throwout lever may be used to disengage the worm drive for making large changes in azimuth rapidly. Turn the telescope, using the orienting knob until the datum point appears at the center of the reticle of the right-hand telescope. The orienting clamping knob may be temporarily released for making large angular changes rapidly. Thereafter, use only the azimuth knob, or, for large changes, the azimuth throwout lever, and the correct azimuth of the point observed will be indicated. For azimuths in the 3,200 to 6,400-mil region, additional numbers (0 to 3,200) mils are provided, corresponding to the azimuth scales on panoramic telescopes and other instruments used by field artillery organizations.

(4) To read angle of site, swing the angle of site mechanism into a substantially vertical plane. Direct the telescope on the object and rotate the elevating knob until the object appears at the center of the reticle. By means of the angle of site knob, center the bubble of the angle of site level in its vial. The angle of site is then read on the angle of site scale (100-mil steps) and micrometer (1-mil steps). An indication of 300 mils corresponds to a horizontal line of sight.

(5) Angular indications are on the reticle. The horizontal axis of the reticle is graduated at 5-mil intervals for 30 mils on each side of the center. The two short lines above the horizontal line are spaced 3 mils apart.

(6) To prepare the instrument for traveling, remove the sunshades and filters, if used, and place them in the pockets of the telescope carrying case. Cover the objectives and eyepieces. With the telescope shanks in a vertical position, press the locking plunger and lift the telescope from the mount. Loosen the telescope clamping knob and swing the elevating mechanism against the right- or left-hand telescope. The instrument will then fit snugly into the blocking of the case. The mount need not be removed from the tripod. Tripod leg clamping levers should not protrude.

d. Tests and Adjustments.

(1) The azimuth micrometer and azimuth scale should read "ZERO" simultaneously. The screw in the end of the micrometer may be tem-

FIRE-CONTROL EQUIPMENT

porarily loosened to permit slipping the micrometer to the desired position.

(2) The angle of site mechanism may be checked by observing a datum point of known angle of site. Small errors may be corrected by temporarily loosening the screw in the end of the knob and slipping the micrometer and knob to the correct position. Should the angle of site scale and micrometer then fail to indicate 3 and 0, respectively, simultaneously, the instrument should be turned in for adjustment by authorized ordnance personnel.

(3) The ball-and-socket joint of the mount should have a snug friction fit when the associated clamping lever is released. Excessive tightness or lost motion may be adjusted by means of the plug in the center of the bottom of the mount. This plug is locked by the concentric retaining plug which must be loosened for adjusting; the retaining ring is tightened securely when adjustment is completed.

e. Care and Preservation.

(1) Refer to paragraph 71 for general instructions pertaining to the care and preservation of instruments.

(2) Exposed moving parts should be oiled occasionally. Interior parts are not to be lubricated by the using arms. Keep excess lubricant, that seeps from the mechanisms, wiped off to prevent accumulation of dust and grit.

TM 9-331
86-89
155-MM HOWITZER M1 AND 155-MM HOWITZER CARRIAGE M1

CHAPTER 4

AMMUNITION

	Paragraph
General	86
Nomenclature	87
Firing tables	88
Classification	89
Identification	90
Care, handling, and preservation	91
Authorized rounds	92
Preparation for firing	93
Projectiles	94
Propelling charges	95
Fuzes	96
Primers	97
Packing	98
Subcaliber ammunition	99
Field report of accidents	100

86. GENERAL.

a. Ammunition for the 155-mm Howitzer M1 is of the separate loading type. For this type of ammunition, the loading of each complete round into the cannon requires three separate operations—those of loading: one, the projectile; two, the propelling charge; and three, the primer. These components, as well as the fuzes for the service projectiles, are shipped separately, the fuze being assembled to the projectile just prior to firing the round. A complete round includes all the components necessary to fire the weapon once.

87. NOMENCLATURE.

a. Standard nomenclature is used here in all references to specific items of issue. Its use for all purposes of record is mandatory.

88. FIRING TABLES.

a. For applicable firing tables, see chapter 9.

89. CLASSIFICATION.

a. Dependent upon the kind of filler, projectiles for the 155-mm Howitzer M1 are classified as high-explosive, chemical, or dummy (drill). The high-explosive projectiles contain a high-explosive bursting charge, whereas the chemical projectiles contain a chemical filler; that is, a chemical agent which produces either a toxic or an irritating physiological effect, a screening smoke, an incendiary action, or a combination

AMMUNITION

of these. Dummy projectiles are completely inert, and are provided for training in loading and handling.

90. IDENTIFICATION.

a. *General.* Ammunition and ammunition components are completely identified by means of painting and marking (including an ammunition lot number). Other essential information is marked on the components, for example: on the projectile, the weight zone and kind of filler; on the propelling charge, the weight of igniter, the charge number of each section, etc. See figures 110 to 118, inclusive, and the following paragraphs. The muzzle velocity may be obtained from firing tables.

b. *Mark or Model.* To identify a particular design, a model designation is assigned at the time the item is classified as an adopted type. This model designation becomes an essential part of the standard nomenclature and is included in the marking on the item. The present system of model designation consists of the letter "M" followed by an arabic numeral. Modifications are indicated by adding the letter "A" and appropriate arabic numeral. Thus "M107A1" indicates the first modification of an item for which the original model designation was "M107." Prior to July 1, 1925, it was the practice to assign mark numbers—that is, the word "Mark," abbreviated "Mk.," followed by a Roman numeral.

c. *Ammunition Lot Number.* When ammunition is manufactured, an ammunition lot number, which becomes an essential part of the marking, is assigned in accordance with pertinent specifications. In the case of separate loading ammunition, a separate lot number is assigned and marked on each of the components—projectile, propelling charge, fuze, and primer—as well as on all packing containers. It is required for all purposes of record, including reports on condition, functioning, and accidents in which the ammunition is involved. To provide for the most uniform functioning, all of the components in any one lot are manufactured under as nearly identical conditions as practicable. For example, in the case of projectiles, any one lot consists of projectiles made by one manufacturer, assembled by one manufacturer, and of one weight zone. Therefore, to obtain the greatest accuracy when firing separate loading ammunition, successive rounds should consist of:

(1) Projectiles of one lot number (one kind and one weight zone).
(2) Propelling charges of one lot number.
(3) Fuzes of one lot number.
(4) Primers of one lot number.

d. *Painting and Marking.*

(1) PAINTING. Projectiles are painted to prevent rust and to provide, by the color, a ready means of identification as to type. The color scheme is as follows:

155-MM HOWITZER M1 AND 155-MM HOWITZER CARRIAGE M1

High-explosive.	Olive-drab, marking in yellow.
Chemical.	Gray, 2 green bands to indicate persistent gas, or one yellow band to indicate smoke; marking is in the same color as the bands.
Dummy or drill (inert).	Black, marking in white, except band at center of gravity which is painted red.

(2) MARKING. For purposes of identification, the following is marked on the components of the service ammunition described herein:

(a) On the Projectile (Stenciled).
Weight zone marking.
Caliber and type of cannon in which fired.
Kind of filler, for example: TNT, HS GAS, etc.
Lot number of filled projectile.[*]
Model of projectile.

(b) Stamped Just Above Rotating Band.
Caliber and designation of the shell.
Lot number of metal parts assembly, year of manufacture, and initials or symbol of manufacturer.

(c) On the Propelling Charge or Sections Thereof (Stenciled on the End).
Model number of charge.
Type of section, that is "BASE" or "INCREMENT."
Number of the charge—"CHARGE 1," "CHARGE 2," etc.
Powder lot number (includes type of powder, the word "LOT," initials of manufacturer, serial number of the lot, and year of manufacture).
Caliber and type of cannon in which fired.
On the igniter end of base section: weight of igniter, type of igniter powder, and caliber and type of cannon in which fired.

(d) On the Fuze (Stamped on the Body).
Type and model of fuze.
Initials of loader.
Month and year loaded.
Lot number of loaded fuze.[*]
Action—"SQ," "DEL"; time, in seconds.

(e) On the Primer (Stamped on the Head).
Initials of loader.
Lot number of loaded primer.[*]
Year of loading.
Mark number of primer.

---

[*] Under revised specifications, future lot numbers will include manufacturer's initials.

AMMUNITION

c. **Weight Zone Markings.** Variations in weight occur during the manufacture of most high explosive and chemical projectiles. In such cases, the projectiles are grouped in weight zones in order that the appropriate ballistic corrections provided in firing tables may be applied for the variations in weight. The weight zone of the projectile is indicated thereon by means of squares with a prick punch in the center of each square—one, two, three, four, or more being used, depending upon the weight of the projectile. For 155-mm projectiles, four squares indicate "normal" or "standard" weight, that is, the weight zone for which no weight corrections are necessary when computing a range of fire.

91. CARE, HANDLING, AND PRESERVATION.

a. **General.** Ammunition components are packed to withstand conditions ordinarily encountered in the field. As shipped, the high explosive and chemical projectiles are provided with grommets to protect the rotating bands, and, in addition, are fitted with eyebolt lifting plugs. Dummy projectiles are crated for shipment. Propelling charges, fuzes, and primers are packed in moisture-resistant containers. Since explosives are adversely affected by moisture and high temperature, due consideration should be given to the following:

(1) Do not break the moisture-resistant seal until ammunition is to be used.

(2) Protect the ammunition, particularly fuzes, primers, and propelling charges, from high temperatures, including the direct rays of the sun. More uniform firing is obtained if the rounds, propelling charges especially, are at the same temperature.

b. Do not remove the eyebolt lifting plug from unfuzed rounds until the fuze is to be assembled thereto. The eyebolt lifting plug is provided for convenience in handling and to keep the fuze opening free of foreign matter.

c. Do not attempt to disassemble any fuze.

d. Do not remove protection or safety devices from fuzes until just before use.

e. Explosive ammunition must be handled with appropriate care at all times. The explosive elements in primers and fuzes are particularly sensitive to undue shock and high temperature.

f. Primers must always be stored in a dry place. Prolonged exposure to moisture or dampness may cause malfunctioning.

g. When it is necessary to leave ammunition in the open, raise it on dunnage at least 6 inches from the ground and cover with a double thickness of paulin. Suitable trenches should be dug to prevent water from flowing under the pile. For further precautions in storage, see TM 9-1900, Ammunition, General.

TM 9-331
91-92

155-MM HOWITZER M1 AND 155-MM HOWITZER CARRIAGE M1

h. Each of the separate loading components should be free of foreign matter—sand, mud, grease, etc.—before loading into the gun.

i. Components of rounds prepared for firing but not fired, will be returned to their original condition and packings, and appropriately marked. Fuzes will be inspected prior to repacking. Such components will be used first in subsequent firing, in order that stocks of opened packings may be kept at a minimum.

j. Do not handle duds. Because their fuzes are armed, duds are extremely dangerous. They will not be moved or turned, but will be destroyed in place in accordance with TM 9-1900.

92. AUTHORIZED ROUNDS.

a. The ammunition authorized for use in the 155-mm Howitzer M1 is listed below. It will be noted that the nomenclature completely identifies the ammunition. The number preceding the projectile nomenclature refers to the assembly number in Table II, paragraph 94.

TABLE I

AMMUNITION FOR 155-MM HOWITZER M1

(In addition to the components shown, one PRIMER, percussion, 21-grain, Mk. IIA1, is required for each complete round[1,2].)

Projectile[3]	Propelling Charge[4]	Fuze Type and Model	Action
	Service Ammunition		
(1) SHELL, H.E., M107, unfuzed, 155-mm how., M1 (adapted for FUZE, P.D., M51, w/BOOSTER, M21, or M51A1, w/BOOSTER, M21A1; or FUZE, time, mechanical, M67, w/BOOSTER, M21A1).	CHARGE, propelling, M3 (green bag), 155-mm how., M1. or CHARGE, propelling, M4 (white bag), 155-mm how., M1.	FUZE, P.D., M51, w/BOOSTER, M21. or FUZE, P.D., M51A1, w/BOOSTER, M21A1. or FUZE, time, mechanical, M67, w/BOOSTER, M21A1.	SQ or DEL SQ or DEL TIME
(2) SHELL, smoke, HC, B.E., M116, unfuzed, 155-mm how., M1 (adapted for FUZE, P.D., M54).	CHARGE, propelling, M3 (green bag), 155-mm how., M1. or CHARGE, propelling, M4 (white bag), 155-mm how., M1.	FUZE, P.D., M54.	TIME or SQ

[1] PRIMER, percussion, 21-grain, Mk. II or Mk. IIA, authorized for use when PRIMER, percussion, 21-grain, Mk. IIA1, is not available.
[2] A fixed service primer is generally used for drill purposes with separate loading ammunition. The live primer issued with each round of separate loading drill ammunition is fired by the service and retained for this purpose.
[3] Fitted with a grommet and an eyebolt lifting plug; fuzes to be assembled in the field.
[4] Green bag charge includes charges 1 to 5, inclusive, and the white bag charge, charges 3, 6, 7 only.

164

AMMUNITION

Projectile[1]	Propelling Charge[4]	Fuze Type and Model	Action
(3) SHELL, gas, persistent, HS, M110, unfuzed, 155-mm how., M1 (adapted for FUZE, P.D., M51, w/BOOSTER, M21, or M51A1, w/BOOSTER, M21A1).	CHARGE, propelling, M3 (green bag), 155-mm how., M1. or CHARGE, propelling, M4 (white bag), 155-mm how., M1.	FUZE, P.D., M51, w/BOOSTER, M21. or FUZE, P.D., M51A1, w/BOOSTER, M21A1.	SQ or DEL SQ or DEL
(4) SHELL, smoke, FS, M110, unfuzed, 155-mm how., M1 (adapted for FUZE, P.D., M51, w/BOOSTER, M21, or M51A1, w/BOOSTER, M21A1).	CHARGE, propelling, M3 (green bag), 155-mm how., M1. or CHARGE, propelling, M4 (white bag), 155-mm how., M1.	FUZE, P.D., M51, w/BOOSTER, M21. or FUZE, P.D., M51A1, w/BOOSTER, M21A1.	SQ or DEL SQ or DEL
(5) SHELL, smoke, phosphorus, WP, M110, unfuzed, 155-mm how., M1 (adapted for FUZE, P.D., M51, w/BOOSTER, M21, or M51A1, w/BOOSTER, M21A1).	CHARGE, propelling, M3 (green bag), 155-mm how., M1. or CHARGE, propelling, M4 (white bag), 155-mm how., M1.	FUZE, P.D., M51, w/BOOSTER, M21. or FUZE, P.D., M51A1, w/BOOSTER, M21A1.	SQ or DEL SQ or DEL
	Drill Ammunition		
(6) PROJECTILE, dummy, 95-lb., Mk. I, 155-mm how.	CHARGE, propelling, dummy (base and 6 increments) M2, 155-mm how., M1917-17A1-18 and M1. or CHARGE, propelling, dummy (base and 6 increments), Mk. I, 155-mm how., M1917-17A1-18 and M1.	FUZE, inert, combination, 45-sec., M1907M.	INERT

B.E.—Base ejection
DEL—Delay
FS—Sulfur trioxide-chlorsulfonic acid solution
HC—Hexachlorethane-zinc mixture
HS—Mustard gas
P.D.—Point-detonating
SQ—Superquick
WP—White phosphorus

[1] Fitted with a grommet and an eyebolt lifting plug; fuzes to be assembled in the field.
[4] Green bag charge includes charges 1 to 5, inclusive, and the white bag charge, charges 5, 6, 7 only.

93. PREPARATION FOR FIRING.

a. The separately issued components are prepared for firing as follows:

b. Projectile. The grommet and eyebolt lifting plug are removed and the appropriate fuze is assembled to the projectile as described in paragraph 94.

c. Propelling Charge. The igniter protector cap is removed and the charge is adjusted for the zone to be fired, as described in paragraph 95.

d. Fuze. Set the fuze as described in the paragraph devoted to the particular fuze.

e. Primer. The primer is ready for firing as shipped, and need only be inserted into the firing mechanism of the cannon.

NOTE: If ammunition components prepared for firing are not fired, return them to their original condition and packing. Reset time fuzes to "SAFE" before storing.

TM 9-331
94

155-MM HOWITZER M1 AND 155-MM HOWITZER CARRIAGE M1

94. PROJECTILES.

a. **General.** Service projectiles for 155-mm howitzers are of two general design types—of an earlier, and a current design. Only projectiles of current design are authorized for use in the 155-mm Howitzer M1. These can be readily distinguished by the marking thereon and by the rotating bands. Bands of the projectiles of current design are uniformly 1.02 inches wide, whereas those on the earlier shell are 0.60 inches wide. It will be noted that there is no shrapnel authorized for the M1 Howitzer.

b. **Service Projectiles.** High-explosive and chemical projectiles authorized for the M1 Howitzer have approximately the same ballistic shape and external dimensions. The projectiles are "boat-tailed" (surface to the rear of the rotating band is conical), have a single rotating band located about 3.5 inches in front of the base, and in all cases have an ogive with a radius of approximately 10.75 calibers. The fuzed shell is approximately 27.5 inches long. However, weights of the loaded projectiles will vary, depending on their fillers (see Table II). A steel base plate welded to the base end of the high explosive projectile prevents gas from the propelling charge from reaching the bursting charge of the projectile through possible flaws in the base. The M110 chemical shell has an adapter in the nose cavity to which is fastened one end of the bursting charge casing. The nose of the M116 chemical shell is in two sections. The forepart contains the bursting charge and the first assembly of smoke filler; and is permanently secured to the main shell casing containing the remaining smoke filler assemblies. The base of the projectile is closed by a base plug which, in the functioning of the shell, is blown out, ejecting the smoke filler.

c. **Preparation for Firing.** To prepare the unfuzed projectiles for firing, remove the grommet and assemble the fuze to the projectile. This is done in the following steps:

(1) Unscrew the eyebolt lifting plug.

(2) Inspect the fuse cavity and threads. They should be free of foreign matter which would interfere with the proper assembly of the fuze.

(3) Remove the cotter pin from the booster of the fuze (except the M54 fuze, which has no booster).

(4) Screw the fuze into the projectile or adapter by hand. Tighten with the fuze wrench.

(5) Set the fuze as described in paragraph 96.

d. **Dummy (Drill) Projectile.** The PROJECTILE, dummy, 95-lb, Mk. I, 155-mm how., is provided for training in the service of the 155-mm gun as well as the howitzer. However, when used with the howitzer the projectile is fitted with a rotating band having a maximum diameter of 6.1 inches—when used in the gun the band has a diameter of 6.5 inches. The principal parts of this dummy projectile are a cast-iron body, a steel

TABLE II
CHARACTERISTICS OF PROJECTILES FOR 155-MM HOWITZER M1

Assembly No.	Figure No.	Kind	Type	Model	Weight As Fired (lb.)	Charge Kind	Charge Weight (lb.)	Length As Shipped (In.)	Length As Fired (In.)	Other Characteristics
SERVICE AMMUNITION										
1	111	SHELL	H.E.	M107	95.00*	TNT†	15.13†	26.81	27.57	Radius of ogive, 10.75 calibers
2	113	SHELL	Smoke (base ejection)	M116	95.10	HC	25.84	27.27	27.27	
3	112	SHELL	Gas, persistent	M110	94.20	HS	11.70	26.78	27.54	
4		SHELL	Smoke	M110	99.40	FS	16.90	26.78	27.54	Single rotating band, 1.02 inches wide
5		SHELL	Smoke	M110	98.09	WP	15.60	26.78	27.54	
DUMMY AMMUNITION										
6		PROJECTILE	Dummy	Mk. I	95.00			21.14	21.14	Radius of ogive, 2 calibers. Single rotating band, 1.19 inches wide

FS—Sulfur trioxide-chlorsulfonic acid solution.
HC—Hexachlorethane and zinc mixture.
H.E.—High explosive.
* Loaded with TNT—see note † below.
† Alternatives: Trimonite, 15.65 lbs.; Amatol. 50-50, 14.19 lbs.; Amatol. 80-20, 13.36 lbs.
WP—White phosphorus.
HS—Mustard gas.
In.—inches. Lb.—pounds.

155-MM HOWITZER M1 AND 155-MM HOWITZER CARRIAGE M1

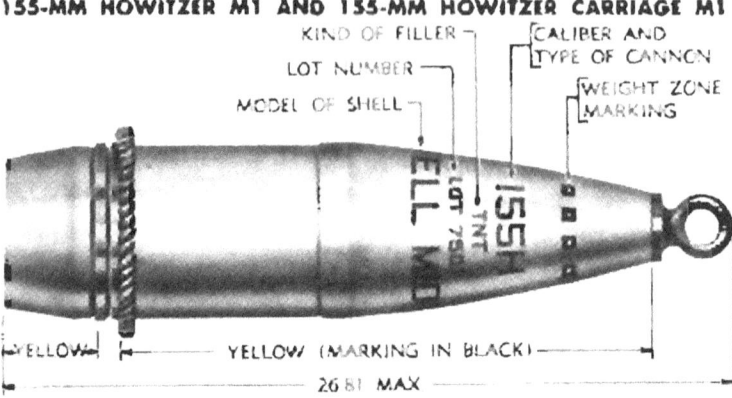

Figure 110—SHELL, H. E. M107, Unfuzed, 155-mm Howitzer M1 (Adapted for FUZE, P. D., M51, w/BOOSTER, M21, M51A1, w/BOOSTER, M21A1, or FUZE, Time, Mechanical, M67, w/BOOSTER, M21A1)

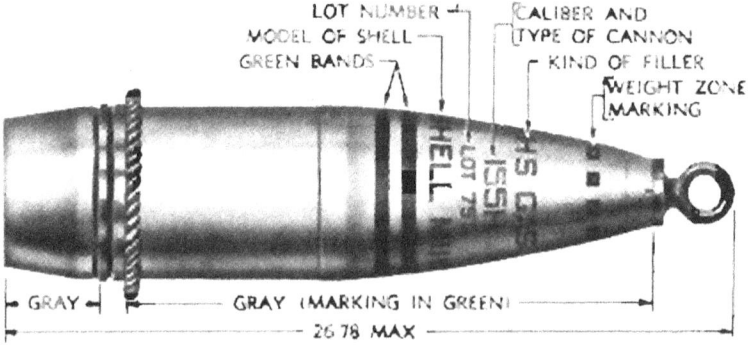

Figure 111—SHELL, Gas, Persistent, HS, M110, Unfuzed, 155-mm Howitzer M1 (Adapted for FUZE, P. D., M51, w/BOOSTER, M21, or M57A1, w/BOOSTER, M21A1)

base, a bronze front band, and a steel rear ring on which is mounted a bronze band. The cast-iron body is ogival and screws on to the steel base at about the center of gravity of the projectile. The bronze front band, fixed to the body by countersunk screws, simulates the bourrelet of a service shell. The bronze rear band simulates the rotating band of a service projectile. The steel ring on which it is mounted slides freely along the cylindrical portion of the steel base. The individual parts are replaceable. As the projectile is not fired, it must be removed from the gun after each loading operation by means of the extractor.

c. Pertinent data on the projectiles are given in Table II.

AMMUNITION

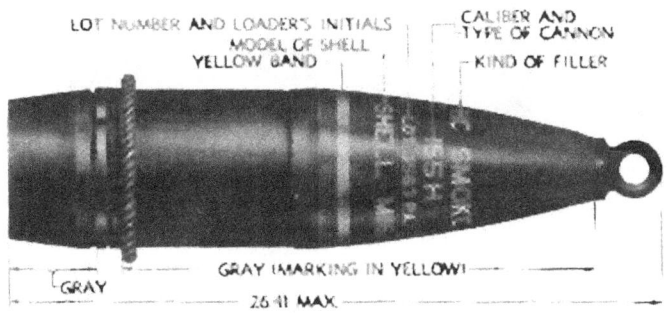

Figure 112 — SHELL, Smoke, HC, B. E., M116, Unfuzed, 155-mm Howitzer M1 (Adapted for FUZE, P. D., M54) As Shipped

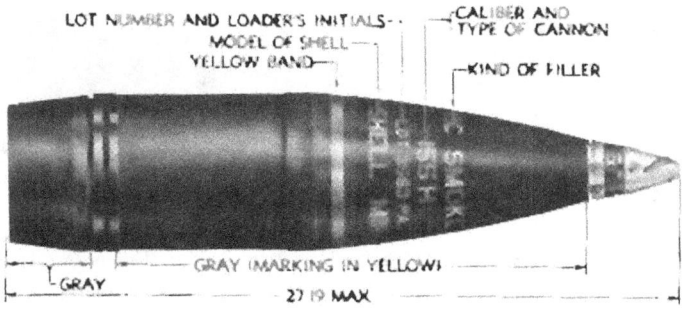

Figure 113 — SHELL, Smoke, HC, B. E., M116, Unfuzed, 155-mm Howitzer M1 (Adapted for FUZE, P. D., M54) As Fired

95. PROPELLING CHARGES.

a. General. The propelling charges for the 155-mm Howitzer M1, are divided into a base section and unequally sized increments, to provide for zone firing. The service propelling charges consist of smokeless powder in cloth bags to which is attached the igniter charge of 3 ounces of black powder, contained in an igniter cloth bag. One 3-ounce igniter pad is sewed to the rear end (breech end) of the base section. The cloth of the igniter pad is dyed red for identification and to indicate the presence of black powder. An igniter protector cap is placed over the exposed igniter to protect it during shipment and storage. Four tying straps sewed to the base section provide a means whereby the increments are attached thereto.

155-MM HOWITZER M1 AND 155-MM HOWITZER CARRIAGE M1

Figure 114 — CHARGE, Propelling, M3 (Green Bag), 155-mm Howitzer M1

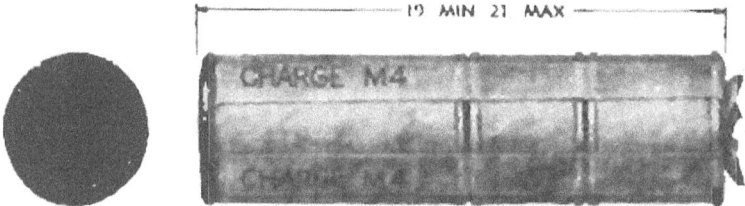

Figure 115 — CHARGE, Propelling, M4 (White Bag), 155-mm Howitzer M1

b. *Types.* There are two types of service charge. One consists of a base section and 4 increments comprising charges 1 to 5 inclusive, contained in cartridge bags of green cloth. It is commonly referred to as the "green bag" charge, to distinguish it from the "white bag" charge which consists of a base section and 2 increments, assembled in cartridge bags of white cloth. The "white bag" charge comprises charges 5 to 7 inclusive. Sections of the green bag charge will never be mixed with sections of the white bag charge, as the latter contains powder which has a slower burning rate. The use of charges having increments of more than one color is, therefore, definitely prohibited. These two charges are similar in general design for corresponding charges for the 155-mm Howitzer M1917-17A1-18, but should not be confused with them as the charges for the two cannons are not interchangeable. The following identifying markings appear on the rear end of the base section:

CHGE. (M3 or M4)
IGNITER
3 OZ. GR. A1
BLK PDR
155 MM. H.

AMMUNITION

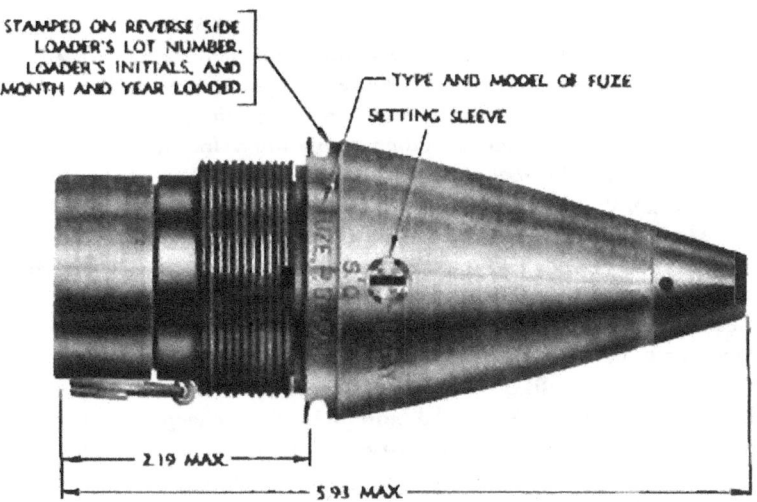

Figure 116—FUZE, P. D., M51A1, w/BOOSTER, M21A1

c. **Preparation for Firing.** When firing at ranges requiring the full charge (Charge 5 for green bag, Charge 7 for white bag), it is only necessary to remove the igniter protector cap prior to loading into the gun. For all other ranges of fire it is also necessary to adjust the propelling charge by removing those increments not required. This is done by untying the straps, removing such increments, and then retying the straps. The charge number is stenciled on the front of each increment. Care should be exercised to keep the increments with the large number uppermost, and in the proper numerical order.

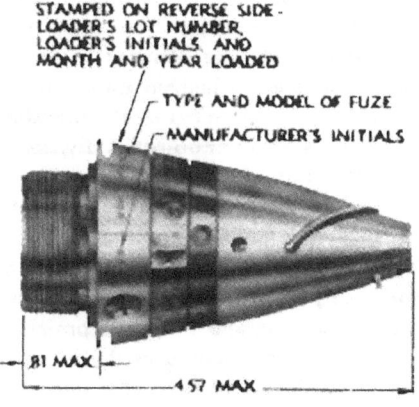

Figure 117—FUZE, P. D., M54

155-MM HOWITZER M1 AND 155-MM HOWITZER CARRIAGE M1

d. Charge, Propelling, M3 (Green Bag), 155-mm How., M1. This charge (fig. 114) consists of a base section and 4 increments. It weighs 5.94 pounds and is authorized for firing charges 1 to 5 inclusive. Its maximum length is 16 inches and the maximum diameter is 5 inches. For additional information, including preparation for firing, see subparagraphs a, b, and c above.

e. Charge, Propelling, M4 (White Bag), 155-mm How., M1. This white bag charge (fig. 116) consists of a base section and 2 increments. It weighs 13.86 pounds and is authorized for firing charges 5 to 7 inclusive. Its maximum length is 21 inches and the maximum diameter is 5.8 inches. For further information, including preparation for firing, see subparagraphs a, b, and c above.

f. Charge, Propelling, Dummy (Base and 6 Increments), M2, 155-mm How., M1917-17A1-18 and M1. This charge simulates a service propelling charge. It has a base and 6 increment sections made up of lead-weighted wood cylinders covered with cotton duck. There is no igniter pad. A cotton-webbed handle is fastened to the rear end of the base section for convenience in handling and extraction. The complete charge is approximately 11 inches long and weighs approximately 7.37 pounds.

g. Charge, Propelling, Dummy (Base and 6 Increments), Mk. I, 155-mm How., M1917-17A1-18 and M1. This dummy propelling charge resembles the M2 dummy charge in general design and contour. However, the over-all length is approximately 12.5 inches, and the nominal weight 8 pounds.

96. FUZES.

a. General. A fuze is a mechanical device used with a projectile to explode it at the time and under the circumstances desired. Fuzes are classified, according to the manner of functioning, as "time" or "impact." Time fuzes contain a graduated time element in the form of a compressed black powder train or a mechanism similar to clockwork which is set to explode the shell a certain number of seconds after firing. Impact fuzes function upon striking a resistant object. A division of impact fuzes, according to the speed of action after impact, is made into "superquick" and "delay." Because of their location on the projectile, the impact fuzes described herein are known as point-detonating (P.D.).

b. Boresafe Fuzes. Certain fuzes are considered to be "boresafe." A boresafe (detonator-safe) fuze is one in which the explosive train is so interrupted that, prior to firing and while the projectile is still in the bore of the cannon, premature detonation of the bursting charge of the projectile is prevented should any of the more sensitive explosive elements in the fuze, primer and/or detonator, malfunction. All fuzes described in this section fall within the boresafe classification.

AMMUNITION

NOTE: Fuzes will not be disassembled. The only authorized assembling or disassembling operation is that of assembling the fuze to the projectile, or, if not fired, removing the fuze from the projectile. Any attempt to disassemble fuzes in the field is dangerous and is prohibited except under specific directions from the Chief of Ordnance.

c. FUZE, P.D., M51A1, w/BOOSTER, M21A1.

(1) DESCRIPTION. In this fuze (fig. 116), the booster, instead of being a component of the loaded projectile, is permanently attached to the fuze at the time of manufacture and thereafter handled and assembled to the projectile as a unit. The fuze contains 2 actions, "superquick" and "delay." Although both actions are initiated on impact, functioning of the shell depends upon the setting of the fuze. When the fuze is set "DELAY," the superquick action is so interrupted that the projectile functions with delay action. It should be noted, however, that if the superquick action malfunctions when the fuze is set "SQ," the projectile will function with delay action rather than be a dud. On the side of the fuze near the base is a slotted "setting sleeve" and 2 registration lines: the one parallel to the axis is marked "SQ," the other "DELAY." As shipped, the fuze is set "SQ." To set the fuze for delay action it is only necessary to turn the setting sleeve so that its slot is alined with "DELAY." A delay pellet — 0.05 second — incorporated in the delay action train provides for the delay action. The setting may be made or changed at will with a screwdriver or other similar instrument any time before firing. This can be done, even in the dark, by noting the position of the slot—parallel to the fuze axis for superquick action, at right angles thereto for delay. A cotter pin with pull ring is assembled to the booster to prevent accidental movement of the detonator during shipment. This cotter pin is to be withdrawn just prior to assembling the fuze with booster to the projectile.

(2) PREPARATION FOR FIRING. Once assembled to the projectile, as outlined in paragraph 94, the fuze is ready for firing except for setting for the required action. If delay action is required, aline slot in setting sleeve with "DELAY"; if superquick, aline slot with "SQ" setting as shipped. Fuze may be reset as required.

d. FUZE, P.D., M51, w/BOOSTER, M21.

(1) DESCRIPTION. This fuze is similar to the FUZE, P.D., M51A1, w/BOOSTER, M21A1, described in subparagraph *c* above, except for minor internal details in the booster.

(2) PREPARATION FOR FIRING. This is the same as described for the FUZE, P.D., M51A1, w/BOOSTER, M21A1.

e. FUZE, P.D., M54.

(1) DESCRIPTION. This fuze is a combination time and superquick type. A safety pull wire extends through the fuze to secure the time

155-MM HOWITZER M1 AND 155-MM HOWITZER CARRIAGE M1

plunger during shipment. The fuze contains 2 actions, time and superquick. The superquick action is always operative and will function on impact unless prior functioning has been caused by time action. Therefore, to set the fuze for superquick action, it is required that the time action be set either at safe (S) or for a time longer than the expected time of flight. The time-train ring, graduated for 25 seconds, is similar to that of the powder time-train fuzes. To prevent extremely short time action, an internal safety feature prevents the time action from functioning should the fuze be set for less than 0.4 second. Therefore, when setting for time action, the setting should always be greater than this minimum of 0.4 second. The fuze as shipped is set safe (S); prior to firing the fuze is set for the required time by means of a fuze setter.

(2) PREPARATION FOR FIRING. Prior to firing, but not before assembly to the projectile, the safety pull wire must be withdrawn from the fuze for either superquick or time setting (pull lower end of the wire from the hole and slide wire off the end of the fuze). If superquick action is required, the graduated time-train ring can be left as shipped (set at safe (S)), or set for a time greater than the expected time of flight. If time action is required, the graduated time-train ring is set for the required time of burning by means of a fuze setter.

NOTE: If, after setting the fuze preparatory to firing, the round is not fired, the fuze will be reset "safe" (S) and the safety pull wire replaced in its proper position, before the round is returned to its packing container.

f. FUZE, Time, Mechanical, M67, w/BOOSTER, M21A1.

(1) DESCRIPTION. This fuze has been designed to provide a means of high burst adjustment at longer ranges when firing the high explosive shell. It replaces the M55 and M55A1 fuzes for that purpose. The fuze is of the mechanical (clockwork) type, and is similar to the M43 type fuze in contour, and in design except that the escapement mechanism has been modified to give a longer burning time. There is no impact element. The upper and lower caps are staked together and turn as a unit when setting the fuze. A set or register line is stamped on the run of the lower cap. A safety line with "S" below it, and time graduations to 75 seconds with 0.5-second intervals, are stamped on the body. The graduations run counterclockwise, viewed from the point of the fuze. Two setting grooves, one each on the lower cap and body, are provided for setting the fuze. A safety feature incorporated in the fuze is designed to prevent functioning should the fuze be set for 3 seconds or less. As shipped, the fuze is set safe, that is, the set line in the lower cap is in alinement with the safety line "S" in the body. A pull wire is fitted to the fuze to secure the firing pin prior to firing. A cotter pin with pull ring is assembled to the booster to prevent accidental movement of the detonator during shipment. The booster is assembled to the fuze at the

AMMUNITION

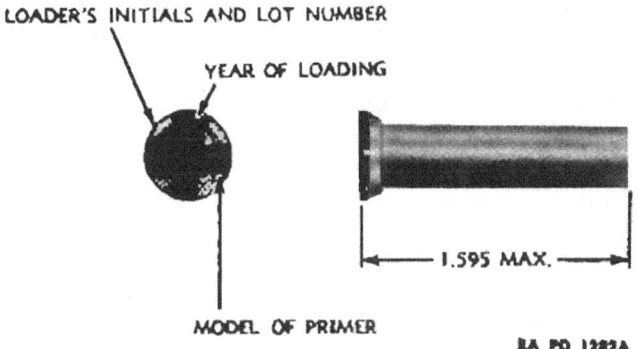

Figure 118—PRIMER, Percussion, 21-grain, Mk. IIA1

time of manufacture and handled thereafter as a single unit with the fuze in shipment and assembly. The cotter pin is to be withdrawn just prior to assembling the fuze with booster to the shell.

(2) PREPARATION FOR FIRING. Once assembled to the projectile, as outlined in paragraph 94, the fuze is prepared for firing by:

(a) Removing the pull wire. This can be done readily by pulling the wire from the hole in the lower cap and sliding the wire off the end of the fuze.

(b) Setting the fuze for the required time by means of the fuze setter, the lower cap being turned counterclockwise as viewed from the point of the fuze. The torque required to set the fuze is between 80 and 100 inch-pounds.

NOTE: Fuzes prepared for firing but not fired will be reset safe (S), and safety devices restored, before returning to packings.

97. PRIMERS.

a. General. A primer used with rounds of separate loading ammunition consists in general of a small quantity of sensitive explosive together with a larger quantity of black powder—all contained in a brass cylindrical container similar in shape to a shot-gun shell or a blank cartridge. The function of the primer is to fire the igniter charge which is attached to the service propelling charge.

b. PRIMER, Percussion, 21-grain, Mk. IIA1.

(1) DESCRIPTION. This primer (fig. 118) is standard for use in the 155-mm Howitzer M1. It consists of a brass case containing a percussion element and 21 grains of black powder. The percussion element in the head of the primer contains a sensitive explosive, hence, should be protected from any blows which might cause accidental functioning.

TM 9-331
97-99

155-MM HOWITZER M1 AND 155-MM HOWITZER CARRIAGE M1

(2) PREPARATION FOR FIRING. To prepare for firing, it is only necessary to insert the primer in the firing mechanism.

c. PRIMER, Percussion, 21-grain, Mk. IIA. This primer is similar to, and is authorized for use in lieu of, the PRIMER, percussion, 21-grain, Mk. IIA1, until present stocks are exhausted.

d. PRIMER, Percussion, 21-grain, Mk. II. This primer is similar to, and is authorized for use in lieu of, the PRIMER, percussion, 21-grain, Mk. IIA1, until present stocks are exhausted.

98. PACKING.

a. Packing. Complete packing data covering dimensions, volume, and weight of the various ammunition components described herein are published in SNL R-2, R-3, and R-6. Service projectiles are shipped uncrated with eyebolt lifting plug and grommet; dummy projectiles, in crates. Propelling charges, fuzes, and primers, are shipped in packings indicated in Table III.

b. Data. Weights of projectiles vary somewhat, dependent upon type and model—propelling charges likewise, dependent upon the particular powder charge. However, the data given in Table III are considered representative for estimating weight and volume requirements.

c. Marking for Shipment. Packings for shipment, and bundle packing identification and shipping plates, are marked as follows:
  (1) Name and address of consignee (or code marking).
  (2) List and description of contents, and A.I.C. symbol.
  (3) Gross weight in pounds, displacement in cubic feet.
  (4) The number of the package.
  (5) The letters "U. S." in several conspicuous places.
  (6) Order number, contract number, or shipping number.
  (7) Ordnance insignium and escutcheon.
  (8) Name or designation of consignor preceded by the word "From."
  (9) Lot number.
  (10) Month and year packed.
  (11) Inspector's stamp.

99. SUBCALIBER AMMUNITION.

a. General. SHELL, fixed, practice, Mk. IIA1, w/FUZE, base, practice, M38, is authorized for use in the 37-mm gun, M1916, when this weapon is used for subcaliber purposes with the 155-mm Howitzer M1. This ammunition is issued in the form of fixed complete rounds. The projectile is fitted with a base fuze, and contains a low-explosive filler of black powder which serves as a spotting charge when the shell is used for target practice. The complete round can be readily identified by the marking thereon.

TABLE III
Packing Data

AMMUNITION

Item	Inner Container	Outer Packing	No. of Items Contained	Weight (pounds)	Volume (cubic feet)	Over-all Dimensions (inches)
SERVICE PROJECTILES	Uncrated, with grommet and lifting plug		1 Proj.	95.00	0.83	26.82 by 7.30 diameter (with rope grommet)
DUMMY PROJECTILE[a]		Crated	1 Proj.	63.00	1.75	$26^{13}/_{16}$ by $10^5/_8$ by $10^5/_8$
PROPELLING CHARGE (M3)	Fiber cntr. (2 charges)	Bundle (3 cntrs.)	6 Charges	53.00	3.00	35.0 by 16.62 by 11.73
	Fiber cntr. (2 charges)	Crate (1 bundle)[aa]	6 Charges	75.00	3.85	$35^3/_8$ by $14^3/_4$ by $12^3/_8$
PROPELLING CHARGE (M4)	Fiber cntr. (1 charge)	Bundle (3 cntrs.)	3 Charges	58.00	2.50	23.75 by 13.97 by 13.04
	Fiber cntr. (1 charge)	Crate (1 bundle)[aa]	3 Charges	79.00	3.93	$27^1/_8$ by 17 by $14^{23}/_{32}$
FUZES	Individual fiber container	Box (25 cntrs.)	25 Fuzes	77.3	1.49	18 by $15^7/_8$ by $9^1/_{32}$
PRIMERS	Metal cntr. (50 primers)	Box (48 cntrs.)	2400 Primers	96.50	1.56	$25^3/_8$ by $10^{13}/_{16}$ by $9^{13}/_{16}$

[a] Shipped complete.
[aa] Overseas shipment.

TM 9-331
99-100
155-MM HOWITZER M1 AND 155-MM HOWITZER CARRIAGE M1

b. Packing.

	Weight (pounds)	Volume (cubic feet)
Complete round without packing material	1.62	
Domestic shipments:		
60 rounds per box w/o metal liner	115	1.60
Over-all dimensions:		
21½ x 12¹¹⁄₁₆ x 10⁵⁄₃₂ inches		
Individual fiber cntrs., 40 cntrs. per box	90.2	1.36
Over-all dimensions,		
18⅛ x 11⅞ x 10²⁹⁄₃₂ inches		
Overseas shipments:		
60 rounds per metal-lined box	128	1.99
Over-all dimensions:		
23⁷⁄₁₆ x 13⁵⁄₁₆ x 11¹⁄₁₆ inches		

100. FIELD REPORT OF ACCIDENTS.

a. Any serious malfunctions of ammunition must be promptly reported to the ordnance officer under whose supervision the material is maintained or issued (par. 7 AR 45-30).

CHAPTER 5

ORGANIZATION SPARE PARTS AND ACCESSORIES

	Paragraph
Organization spare parts	101
Accessories	102

101. ORGANIZATION SPARE PARTS.

a. A set of spare parts is supplied with the 155-mm howitzer and carriage for field replacement of those parts most likely to become broken, worn or otherwise unserviceable. The set should be kept complete at all times by requisitioning new parts for those used. For listing of organization spare parts for the 155-mm howitzer and carriage see SNL C-39.

b. Care of spare parts is covered in the section of this manual entitled "Care and Preservation."

102. ACCESSORIES.

a. Accessories include the tools and equipment required for such disassembling and assembling as the using arms are authorized to perform, and for cleaning and preserving the 155-mm howitzer and carriage. Accessories should not be used for purposes other than those prescribed and when not in use should be properly stored.

b. There are a number of accessories, the names or general characteristics of which indicate their use. Others embodying special features or having special uses, are described in the following paragraphs:

(1) VENT CLEANING BIT. The vent cleaning bit is a steel rod with a looped handle and a fluted end for cleaning the obturator spindle vent.

(2) ARTILLERY GUN BOOK. The Artillery Gun Book (O.O. Form 5825) is used for the purpose of keeping an accurate record of the materiel. It must always remain with the materiel regardless of where it may be sent. The book is divided as follows: Record of assignment; battery commander's daily gun record; inspector's record of examination; forms to be filled out in case of premature explosions. This book should be in the possession of the organization at all times, and its completeness of records and its whereabouts are the responsibility of the battery commander. It must also contain date of issuance of the materiel, by whom used, and the place where issued. If a new howitzer is installed on the carriage, all data recorded in the old book with reference to sights, mounts, etc., must be copied into the new book before the old book is relinquished.

NOTE: Record of assignment data must be removed and destroyed prior to entering combat.

(3) BORE BRUSH, M13. The bore brush is used for cleaning and oil-

TM 9-331
102

155-MM HOWITZER M1 AND 155-MM HOWITZER CARRIAGE M1

ing the bore of the howitzer. In the latter case it is covered with burlap. A six-sectioned wooden staff is used with this bore brush.

(4) HANDSPIKE. The handspike is made of seamless steel tubing and is 5 feet long. It is used to traverse the howitzer by moving the trail.

(5) BLACKOUT LIGHT-SYSTEM. The blackout light-system is strapped around the muzzle of the howitzer for use as a tail lamp when the howitzer is being towed. It has both a red taillight and an amber stoplight for connection to the brake system.

(6) OIL PUMP, M3. The oil pump is provided to fill the recoil and counterrecoil systems. It is a high-pressure type pump operated by a handle coming out of the top of the pump. The filling tube is connected to the pump outlet valve and the other end of the tube is connected to the recoil system. When the pump is being filled, care must be taken that no foreign matter enters the oil reservoir.

(7) CLEANING AND UNLOADING RAMMER, M7. This rammer is made of bronze so that it will not injure the bore of the howitzer. It is cone-shaped to fit the nose of the projectile. It is affixed to the same staff as the bore brush. It is inserted from the muzzle end of the howitzer in order to unload the howitzer or to remove lightly stuck projectiles.

(8) LOADING RAMMER. This is a loading rammer used to ram the projectile into its seat. The rammer is made of bronze and has a wooden staff screwed into it.

(9) PRIMER SEAT CLEANING REAMER. This reamer is used to keep the primer seat free of powder residue. Care must be taken not to injure the primer seat or the reamer.

(10) OIL RELEASE. The oil release is a 5-inch plug-shaped device with a hollow center. Its threaded end is screwed into the filling and drain-plug hole of the recuperator cylinder head for drawing the reserve oil from the recoil mechanism.

(11) LOADING TRAY. The loading tray is used to support the projectile when loading the howitzer. The tray is a concave metal trough with handles attached to the sides for lifting.

(12) FILLING TUBE. The filling tube is made of copper tubing and has a connection at both ends. The tube is 10 feet long so that the oil pump can rest on the ground while the other end of the tube is connected to the filling valve of the recoil mechanism.

(13) FIRING MECHANISM WRENCH. This wrench is used to unscrew the firing pin housing during disassembly of the firing mechanism.

(14) FUZE WRENCH M7. This wrench is used to tighten the fuze in the projectile before firing and for interchanging fuzes. The screwdriver portion of the wrench is used for setting the fuze to either "Superquick" or "Delayed."

(15) SOCKET WRENCH. The socket wrench is used on the wheel stud nuts. It is furnished with a bar handle.

CHAPTER 6

STORAGE AND SHIPMENT

	Paragraph
General instructions	103
Methods of securing howitzer carriage on freight cars	104
Shipment by water	105
Storage	106

103. GENERAL INSTRUCTIONS.

a. *General.* The 155-mm Howitzer Carriage M1 shall be stored and shipped either with or without the howitzer mounted. All precautions should be taken to prevent corrosion during storage or shipment, to keep recoil mechanisms exercised, and to prevent deterioration of rubber during storage.

b. *Preparation.*

(1) LUBRICATION. The materiel should be completely lubricated before storage or shipment.

(2) UNPAINTED SURFACES. All unpainted surfaces should be treated with rust preventives before the carriage is stored or shipped. All such surfaces must be made free from films of moisture, dirt, and other foreign substances. This is accomplished by cleaning the surfaces with a solvent such as SOLVENT, dry-cleaning, or with a soap solution. When a soap solution is used, the surfaces must be rinsed with clean hot water. Before application of corrosion preventives, the surfaces must be thoroughly dried.

(a) Exterior Surfaces. Exterior surfaces shall be treated with COMPOUND, rust-preventive, lead base. This preventive may be applied cold by spraying or brushing.

(b) Interior Surfaces. Interior surfaces from which it would be difficult to remove corrosion preventives, such as the bore of the howitzer, and moving parts in the breech mechanism, shall be treated with a film of COMPOUND, rust-preventive, light. This preventive may be applied by brushing or slushing.

104. METHODS OF SECURING HOWITZER CARRIAGES ON FREIGHT CARS.

a. *Brake Wheel Clearance.* There shall be a 6-inch clearance in back, on both sides of, and above the brake wheel of the flat car, and this clearance should be increased as much as is consistent with proper location of load (A, fig. 119).

b. *Loading.*

(1) INSPECTION. All railroad cars used for shipping the howitzer car-

155-MM HOWITZER M1 AND 155-MM HOWITZER CARRIAGE M1

riage should be inspected to make sure that all floors are sound and that no nails and other projections are present.

(2) RAMPS. Permanent ramps should be used for loading the carriage when available but when such ramps are not available, improvised ramps may be constructed of rail ties and other available lumber.

(3) DISTRIBUTION OF LOAD. The load must be distributed so that there will be, as near as possible, equal weight bearing on each truck of the railroad car.

(4) BRAKES. After loading and placing the carriages, the hand brakes should be set.

(5) PLACARDING. Each railroad car containing the howitzer carriages must be placarded "*DO NOT HUMP.*"

(6) TYPE OF CAR. Box cars, flat cars, or gondola cars may be used.

e. Method of Securing Howitzer and Carriage on Freight Car. (All reference letters in the following paragraphs refer to the details and locations in fig. 119.) There are two approved methods of blocking the 155-mm howitzer carriage as described below.

(1) METHOD 1 (fig. 119).

(a) Blocks D. Four blocks D will be located, one to the front and one to the rear of each wheel. These blocks will have the heel of the block nailed to the car floor with five 40-penny nails and that portion of the block under the wheel toe-nailed to the car floor with two 40-penny nails.

(b) Cleats E. Four cleats E will be located, two against the outside face of each wheel. The lower cleat will be nailed to the car floor with three 40-penny nails, and the top cleats will be nailed to the cleats below with three 40-penny nails.

(c) Supports H. Two supports H will be located under the axle near the inside face of each wheel. These supports will be cut so as to fit snugly between the car floor and the axle. Each support will be nailed to the car floor with four 40-penny nails.

(d) Blocks B. The trails of the carriage will be blocked with five blocks B located as follows:

 1. One against the outside face of each trail.
 2. Two blocks against each end of the trail on each side of the pintle.
 3. One block nailed crosswise to the blocks in subparagraph 2 above, over the pintle. These blocks should be nailed to the car floor or to cleats below with three 40-penny nails.

(e) Strapping C. The trails of the carriage will be strapped with 4 strands, 2 wrappings of No. 8 gage, black annealed wire wrapped once around each trail and passed through the stake pockets. The wires are tightened enough to remove slack. When a box car is used, this strapping should be applied in a similar manner and attached to the floor by use of blocking.

STORAGE AND SHIPMENT

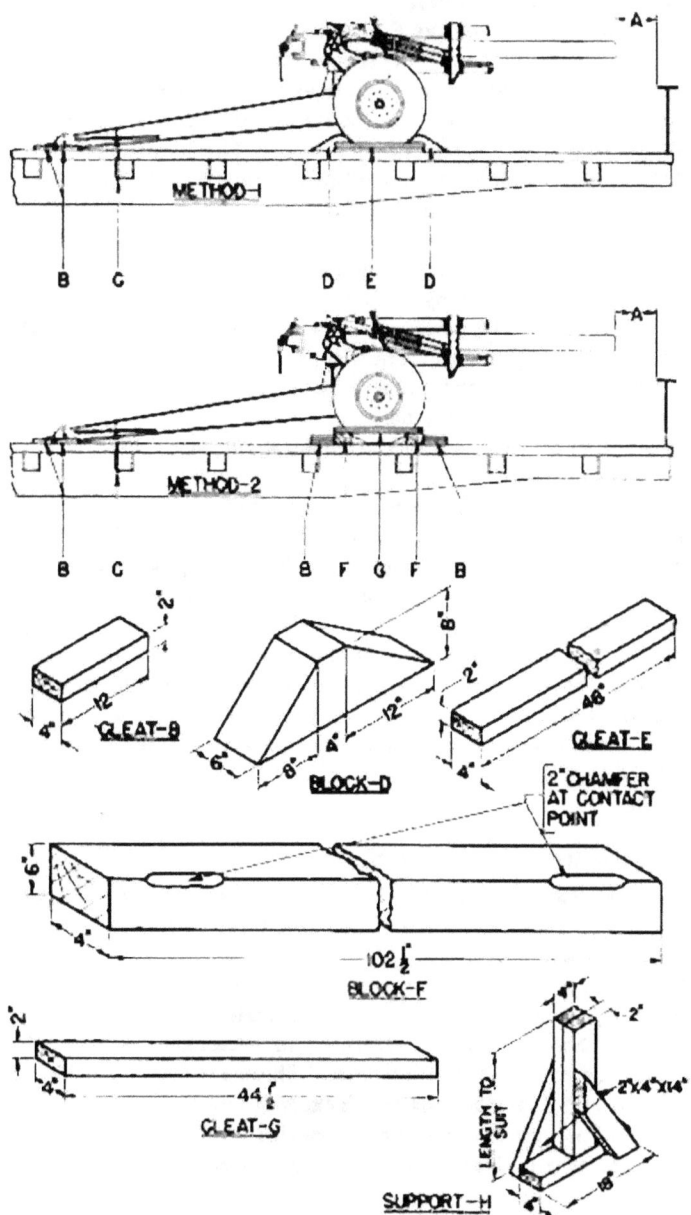

Figure 119 — Blocking Requirements for the 155-mm Howitzer Carriage M1

TM 9-331
104-106

155-MM HOWITZER M1 AND 155-MM HOWITZER CARRIAGE M1

(2) METHOD 2 (fig. 119).

(a) Blocks F. Two blocks *F* shall be placed, one to the front and one to the rear of the wheels. These blocks should be at least 8 inches wider than the over-all width of the vehicle at the car floor.

(b) Cleats B. Eight cleats *B* should be located, two against blocks *F* to the front and to the rear of each wheel. The lower cleats should be nailed to the car floor with three 40-penny nails and the top cleats to the cleats below with three 40-penny nails.

(c) Cleats G. Two cleats *G* will be located, one against the outside face of each wheel on the top of blocks *F*. These cleats shall be nailed to blocks *F* with two 40-penny nails at each end of the cleats.

(d) Supports D. Two supports *D* will be located under the axle near the inside face of each wheel. These supports will be cut so as to fit snugly between the car floor and the axle. Each support will be nailed to the car floor with four 40-penny nails.

(e) Blocks B. The trails of the carriage will be blocked with 5 blocks *B* located as follows:

1. One against the outside face of each trail.

2. Two blocks against each end of the trail on each side of the pintle.

3. One block nailed crosswise to the blocks in subparagraph 2 above, over the pintle. These blocks should be nailed to the car floor or to cleats below with three 40-penny nails.

(f) Strapping C. The trails of the carriage will be strapped with 4 strands, 2 wrappings of No. 8 gage, black annealed wire wrapped once around each trail and passed through the stake pockets. The wires are tightened enough to remove slack. When a box car is used, this strapping should be applied in a similar manner and attached to the floor by use of blocking.

105. SHIPMENT BY WATER.

a. Preparation. The howitzer and carriage will be prepared for overseas shipment as directed in paragraph 103 with special attention to the rust-proofing operations.

b. Crating. In order to protect the materiel and to conserve space aboard ship, it will often be desirable to crate and box the components. Crating or boxing will be performed in accordance with IOSSC-(a), Introduction to Ordnance Storage and Shipment Chart, Section (a), Instructions and Specifications for Packaging Ordnance General Supplies, and should be similar to the crates described in OSSC-C, Ordnance Storage and Shipment Chart, Group C—Major Items.

106. STORAGE.

a. Preparation. The howitzer and carriage will be prepared for storage as directed in paragraph 103.

STORAGE AND SHIPMENT

b. **Rubber Preservation.** When the howitzer and carriage are stored for an indefinite period of time, the tires should be either removed from the wheels or the wheels should be jacked from the ground. If the tires are left mounted on the wheels, the pressure should be reduced 10 pounds per square inch. The tires should then be covered with canvas or some light resistant material. When the tires are removed from the wheels, they should be stored in a dark, dry warehouse.

c. **Exercising the Recoil Mechanism.** During storage, the recoil mechanism and the equilibrators should be exercised at least once every 3 months. During this exercising the moving parts should be inspected for corrosion.

d. **Inspections.** Periodical inspections should be made of the materiel while in storage. This inspection should note among other things, condition of the rust preventives, the missing parts, and the need for repairs. If the carriage is found to be corroding at any part, the entire carriage should be rust-proofed as described in paragraph 103.

TM 9-331
107-108

155-MM HOWITZER M1 AND 155-MM HOWITZER CARRIAGE M1

CHAPTER 7

OPERATION UNDER UNUSUAL CONDITIONS

	Paragraph
General	107
Tropical climates	108
Arctic climates	109
Excessively moist or salty atmosphere	110
Excessively sandy or dusty conditions	111

107. GENERAL.

a. Because of the different climates in which this materiel may be expected to operate, special instructions are given in this section for three regions, namely: arctic, temperate, and tropical.

b. By "arctic" is meant a climate usually experienced in Alaska, Newfoundland, Labrador, or Iceland. By "temperate" is meant a climate usually experienced in continental United States or Hawaii. By "tropic" is meant a climate usually experienced in Panama, the Philippines, or Cuba.

c. In certain cases, the prescribed instructions may not apply; for example, a tropic climate may be experienced in a temperate region. In cases of this nature, the instructions as to the classification of the climate in which the materiel is operating is left to the judgment of the ordnance officer. He is cautioned, however, that only extended, and not temporary, periods of climatic conditions govern the classification.

d. Manufacturing arsenals and plants should lubricate the materiel on assembly as prescribed in the lubrication guides (figs. 50 and 51). If the materiel is to be used in a climate other than temperate, the precautions in paragraphs 108 or 109 below should be taken.

e. Materiel, previously lubricated for a colder climate than the one in which the materiel is to be used, should be relubricated with lubricants prescribed for use in that climate.

f. Materiel, previously lubricated for a warmer climate than the one in which the materiel is to be used, should be completely cleaned of all lubricants and relubricated with the lubricants prescribed for use in that climate.

108. TROPICAL CLIMATES.

a. No definite information is available at this writing other than the usual precautions to see that in temperatures above 90 F summer grade grease (GREASE, O.D., No. 0), and OIL, lubricating, engine, SAE 30, or SAE 50 are used as lubricants. The tires should be checked frequently during traveling to make sure that the pressure is not deviating appre-

OPERATION UNDER UNUSUAL CONDITIONS

ciably from the prescribed pressure. The oil reserves in the replenisher and recuperator and oil levels in gear cases should be checked.

NOTE: Extreme heat is often accompanied by other adverse conditions. Refer to paragraphs 110 and 111 below.

109. ARCTIC CLIMATES.

a. *General.* Preparing a weapon for arctic climates consists of inspecting and placing the weapon in good mechanical condition, cleaning and lubricating with cold weather lubricants, and frequent exercising. The recoil oil should be modified by ordnance maintenance personnel.

b. *Inspection.* The materiel should be inspected to see that all moving parts operate freely and without binding. The elevating and traversing mechanisms should be operated throughout their ranges. Tires should be checked frequently during traveling to see that the pressure is correct. Oil reserves in the replenisher and recuperator should be checked.

c. *Lubrication of the Materiel.* The materiel should be properly lubricated with cold weather lubricants as prescribed in the lubrication guides (figs. 50 and 51). Before applying the cold weather lubricants, the materiel should be thoroughly cleaned and all old lubricants removed. For lubrication and service below 0 F, refer to OFSB 6-S.

d. *Sighting and Fire Control Instruments.* Sighting and fire control instruments are normally lubricated for operation over a wide range of temperatures (including arctic). They should be exercised frequently during periods of low temperature to insure their proper functioning. If the instruments do not function properly, the ordnance maintenance personnel should be notified.

110. EXCESSIVELY MOIST OR SALTY ATMOSPHERE.

a. When the materiel is not in active use, the unpainted parts should be covered with a film of oil or rust-preventive compound. The bore and recoil slide of the tube and the breech mechanism should be kept heavily oiled, and should be inspected daily for traces of the formation of rust. The materiel should be lubricated more frequently than is prescribed for normal service. The breech and muzzle covers must be kept in place as much of the time as firing conditions permit.

b. In excessively salty atmosphere, the oil or rust-preventive compound used should be changed often as the salt has a tendency to emulsify the oil and destroy its rust preventive qualities.

111. EXCESSIVELY SANDY OR DUSTY CONDITIONS.

a. If considerable dust is present when the piece is to be operated, the lubricant should be removed from the elevating and traversing arcs and pinions and the firing jack rack plunger and pinion, and the teeth of these

155-MM HOWITZER M1 AND 155-MM HOWITZER CARRIAGE M1

parts should remain dry until the action is over. If the surfaces are dry, there will be less wear than when coated with a lubricant contaminated with grit.

b. The recoil slide must be kept as free of grit and dirt as is possible. The breather holes in the rear of the replenisher must be kept open. The breech and muzzle covers must be kept in place as much of the time as firing conditions permit.

NOTE: For extreme conditions of speed, heat, water, mud, snow, rough roads, and salty or moist air, lubricate more frequently than is prescribed for normal service.

CHAPTER 8

MATERIEL AFFECTED BY CHEMICALS

	Paragraph
General	112
Protective measures	113
Decontamination	114

112. GENERAL.

a. Gas clouds, chemical shell, and chemical spray are the major chemical warfare methods for destroying or damaging materiel. Removing or destroying the dangerous liquid or solid chemical agents spread by these methods, or changing these chemical agents to harmless substances is called decontamination.

113. PROTECTIVE MEASURES.

a. When materiel (except ammunition) is in constant danger of attack with chemicals, apply a light coat of engine oil to unpainted metal parts. Take care that the oil does not touch the optical parts of instruments, or leather or canvas fittings. Protect materiel not in use with covers as far as possible. Keep ammunition in sealed containers.

b. Ordinary fabrics offer practically no protection against mustard gas or lewisite. Rubber and oilcloth, for example, will be penetrated within a short time. The longer the period of exposure, the greater the danger, when apparel made of either of these materials is worn. Rubber boots contaminated with mustard gas may offer a grave danger to men who wear them several days after the attack. Impermeable clothing, designed to prevent penetration of chemicals, will resist penetration almost indefinitely, but the maximum time such clothing can be worn is from 5 to 10 minutes in summer and about 30 minutes in winter.

114. DECONTAMINATION.

a. For the removal of liquid vesicants (mustard, lewisite, etc.) from materiel, the following steps should be taken:

(1) PROTECTION OF PERSONNEL. For all of these operations a service gas mask and a complete suit of protective clothing, either permeable or impermeable, depending upon the type of contamination, must be worn. Immediately after removing the suit, a thorough bath with soap and water (preferably hot) must be taken. If any skin areas have come in contact with liquid or vapor mustard gas, or if the vapor of mustard has been inhaled, it is imperative that complete first-aid measures be given within 5 minutes to be effective as a preventive. First aid must be prompt, for little can be done later than 20 to 30 minutes after exposure.

155-MM HOWITZER M1 AND 155-MM HOWITZER CARRIAGE M1

(2) Casualties caused by vesicants (mustard, lewisite, etc.) or by lung irritants (phosgene, all vesicants, etc.) should be immediately removed from the contaminated area.

(a) Vesicant Casualties. Remove the contaminated clothing. If the face has been exposed, wash the eyes and rinse the nose and throat with a saturated boric acid, weak sodium bicarbonate, or common salt solution. Mustard burns or skin areas wet with liquid mustard should be immediately and repeatedly swabbed with a solvent, such as kerosene, any oil, alcohol, or CARBON TETRACHLORIDE. Then wash thoroughly with soap and water.

(b) Lung Irritant Casualties. To reduce his oxygen requirements, make the casualty lie down. Keep him warm and give him non-alcoholic stimulants such as hot coffee or tea. He should be evacuated as soon as possible as an absolute litter case.

(c) Complete first aid instructions to supplement the above general instructions are contained in FM 21-40.

(3) Decontaminate garments exposed to vesicants. If impermeable clothing has been exposed to vapor only, it may be decontaminated by hanging in the open air, preferably in sunlight, for several days. It may also be cleaned by steaming for 2 hours. If impermeable clothing has been contaminated with liquid vesicant gases, steam it for 6 to 8 hours. Various kinds of steaming devices can be improvised from equipment in the field.

b. Procedure.

(1) Commence by freeing materiel of dirt through the use of sticks, rags, etc. Sticks, rags and other cleaning items used in decontamination must be burned or buried immediately after their use.

(2) If the surface of the materiel is coated with grease or heavy oil, remove it before decontamination is begun. For this cleaning use SOLVENT, dry-cleaning, or other available solvents for oil, applied on rags attached to the ends of sticks.

(3) Decontaminate the painted surfaces of the materiel with bleaching mixture made by mixing equal parts by weight of AGENT, decontaminating (chloride of lime), and water. So large a proportion of bleaching powder is added to the water that only a small part is dissolved; therefore a suspension, or "slurry" is formed. This slurry should be swabbed over all surfaces. Wash off thoroughly with water, then dry and oil all surfaces.

(4) All unpainted metal parts of materiel that have been exposed to any gas except mustard and lewisite must be cleaned as soon as possible with SOLVENT, dry-cleaning, or ALCOHOL, denatured, and wiped dry. All parts should then be coated with oil.

(5) All unpainted metal parts and instruments exposed to mustard or lewisite must be decontaminated with AGENT, decontaminating, non-

TM 9-331
114

MATERIEL AFFECTED BY CHEMICALS

corrosive, mixed one part solid to fifteen parts solvent (ACETYLENE TETRACHLORIDE) by weight. If this is not available, use warm water and soap. Bleaching slurry must not be used, because of its corrosive action on unpainted metal parts. After decontamination, wipe all metal surfaces dry and coat them lightly with engine oil, except the surfaces of small arms, which must be coated with OIL, lubricating, preservative, light. Instrument lenses may be cleaned only with PAPER, lens, tissue, using a small amount of ALCOHOL, ethyl.

(6) If AGENT, decontaminating (chloride of lime) is not available, materiel may be temporarily cleaned with large volumes of hot water. However, mustard gas lying in joints or in leather or canvas webbing is not removed by this procedure and will remain a constant source of danger until the materiel can be properly decontaminated. Because all mustard gas washed from materiel lies unchanged on the ground, the area should be plainly marked with warning signs before abandonment.

(7) Leather or canvas webbing that has been contaminated should be scrubbed thoroughly with bleaching slurry. If this treatment is believed insufficient, it may be necessary to burn or bury such material.

(8) Ammunition which has been exposed to vesicant gas must be thoroughly cleaned before firing. To clean ammunition use AGENT, decontaminating, noncorrosive, or if this is not available, strong soap and warm water. After cleaning, wipe all ammunition dry with clean rags. *Do not use dry powdered AGENT, decontaminating (chloride of lime) (used for decontaminating certain types of material on or near ammunition supplies), as flaming occurs when it touches liquid mustard.*

(9) **Detailed** information on decontamination is contained in FM 21-40 and TM 3-220.

TM 9-331
115

155-MM HOWITZER M1 AND 155-MM HOWITZER CARRIAGE M1

CHAPTER 9

REFERENCES

	Paragraph
Standard nomenclature lists	115
Explanatory publications	116
Firing tables	117

115. STANDARD NOMENCLATURE LISTS.

   a. **Ammunition.**

Ammunition, fixed and semifixed, all types, for pack, light and medium field artillery, including complete round data	SNL R-1
Ammunition instruction material for pack, light and medium field artillery	SNL R-6
Projectiles and propelling charges, separate loading, all types, for medium field artillery, including complete round data	SNL R-2
Service fuzes and primers for pack, light and medium field artillery	SNL R-3

   b. Cleaning, preserving and lubricating materials; recoil fluids, special oils, and miscellaneous related items   SNL K-1

   c. Firing tables and trajectory charts   SNL F-69

   d. **Gun Materiel.**

Howitzer, 155-mm, M1; and carriage, howitzer, 155-mm, M1	SNL C-39
Major items of pack, light and medium field artillery and armament of these calibers for airplane and combat vehicles	SNL C-1

   e. **Sighting and Fire-Control Equipment.**

Circle, aiming, M1	SNL F-160
Compass, M2	SNL F-219
Finder, range, 1-meter base, M1916	SNL F-26
Light, aiming post, M14	SNL F-220
Light, instrument, M16	SNL F-205
Mount, telescope, M25	SNL F-216
Posts, aiming, M1	SNL F-35
Quadrant, gunner's, M1	SNL F-140
Quadrant, gunner's, M1918	SNL F-13
Sights, bore (small arms and field artillery)	SNL F-10

TM 9-331
115-117

REFERENCES

Tables, firing, graphical	SNL F-237
Telescope, B. C., M1915A1	SNL F-9
Telescope, panoramic, M12	SNL F-214
Current Standard Nomenclature Lists are as tabulated here. An up-to-date list of SNL's is maintained as the "Ordnance Publications for Supply Index"	OPSI

116. **EXPLANATORY PUBLICATIONS.**

   a. Ammunition, general — TM 9-1900

   b. Army Regulations.
      Ordnance field service in time of peace — AR 45-30
      Qualifications in arms and ammunition training allowances — AR 775-10
      Range regulations for firing ammunition for training and target practice — AR 750-10

   c. Auxiliary fire-control instruments (field glasses, eyeglasses, telescopes and watches) — TM 9-575

   d. Cleaning, preserving, lubricating and welding materials and similar items issued by the Ordnance Department — TM 9-850

   e. Maintenance and Repair.
      Artillery lubrication, general — OFSB 6-4
      Cold weather lubrication and service of artillery equipment — OFSB 6-5
      Defense against chemical attack — FM 21-40
      General instructions for recoil fluid, light and medium artillery — OFSB 6-6
      Product guide — OFSB 6-2
      Maintenance of materiel in the hands of troops — OFSB 4-1

   f. Miscellaneous.
      List of publications for training — FM 21-6
      Ordnance provision system regulations — OPSR

   g. Ordnance Storage and Shipment Charts.
      Instructions and specifications for packaging ordnance general supplies — IOSSC-a
      Ordnance storage and shipment chart—group C—major items — OSSC-C

117. **FIRING TABLES.**

   a. Howitzer, 155-mm, M1; Carriage, Howitzer, 155-mm, M1, Shell, H. E., M107 — FT 155-Q-1
   Current firing tables are as tabulated here. An up-to-date list of firing tables is maintained in — SNL F-69

TM 9-331

155-MM HOWITZER M1 AND 155-MM HOWITZER CARRIAGE M1

INDEX

A

	Page No.
Abrasives	89
Accessories	
Artillery Gun Book	179
blackout light-system	180
bore brush, M13	179–180
care of	79
cleaning and unloading rammer, M7	180
compass M2	144
filling tube	180
firing mechanism wrench	180
fuze wrench M7	180
general discussion of	179
handspike	180
loading rammer	180
loading tray	180
oil pump, M3	180
oil release	180
primer seat cleaning reamer	180
socket wrench	180
vent cleaning bit	179
Accidents, field report of ammunition	178
Adapter, description and functioning	21
Adjustments:	
B.C. telescope M1915 or M1915A1	158–159
compass M2	148
equilibrators	44
one-meter base range finder M1916	152–154
sighting equipment	125–126
(See also Inspection and adjustment)	
Aiming circle M1	
care and preservation	139
description	135
instrument light	137
operation	137–139
test and adjustment	139
Aiming post light M14	127–128
Aiming post M1, description	126–127
Air in replenisher or recoil system	67
Alinement	
equilibrators	43–44
howitzer in mount	12
Ammunition	
authorized rounds, table	164–165

	Page No.
care, handling, and preservation	163–164
classification	160–161
cleaning exposed to vesicant gas	191
field report of accidents	178
firing, preparation for	165
firing tables	193
fuzes	172–175
general discussion of	160
identification	161–163
nomenclature	160
packing	176, 177
primers	175–176
projectiles	166–168
characteristics of, table	167
propelling charges	169–172
subcaliber	176
packing	178
Ammunition lot number	161
Angles of sight	
measuring	147
reading	151
Arctic climates, operation in	187
Armor plate shields	4
Artillery Gun Book, use of	179
Assembly:	
block rotating roller	116
breech mechanism	111–115
firing mechanism M1	102–103
general discussion of	100–101
obturating parts	13–14
operating lever latch	115
Authorized rounds, table	164–165
Axle, description	50
Axle and hubs	50–52
Azimuths	
measuring	141
angles	145–147
movement in	7
Azimuth indication	
adjustment	153
aiming circle M1	135

B

Ball check valve, function	36
B.C. telescope M1915 or M1915A1	
care and preservation	159
description	156–157
general discussion of	156
operation	157–158
tests and adjustments	158–159

INDEX

B—Cont'd

	Page No.
Barrel assembly	
breech ring	12
tube	9-12
Battery box	49-50
Bearings and wheels, care of	88
Blackout light-system	180
Block rotating cam	17
Block rotating roller	
assembly	116
disassembly	115
Blocks	182, 184
Bore	
cleaning	58
after firing	79
height of	7
length of	4
oiling	80
Bore brush, M13	179-180
Bore sight, use of	131-132
Bore sighting, procedure for	132-134
Boresafe fuzes	172-173
Bottom carriage	44
pintle	45
pintle bearing	44-45
trails and spades, inspection	98-99
Boxing material for shipment	184
Brake "drag"	63
Brake wheel clearance in railroad car	181
Brakes	
grabbing	71
none or intermittent	70
operation	63
weak	70-71
Breech	
closing	27, 55
after loading howitzer	60
opening	24-27, 55
will not open or fully close with firing mechanism in place	66
Breech mechanism	
assembly	111-115
cleaning after firing	80
cleaning and lubricating	75
description and functioning	12-20
breechblock	12-13
breechblock actuating mechanism	15-17
breechblock carrier	15
counterbalance	17-18
hinge pin	17
obturator	13-15

	Page No.
operating lever latch	18
safety latch	18-20
disassembly	103-111
does not operate freely functioning	66
closing the breech	27
general description	24
opening the breech	24-27
operating	55
servicing	81
Breech races and breechblock threads, inspection	96
Breech ring	
description	12
marking	9
threaded sectors do not mate	66
Breechblock	
description	9
description and functioning	12-13
failure	66
installing, illustration	112
removing, illustration	108
threaded sectors do not mate	66
type of	4
Breechblock actuating mechanism, description and functioning	15-17
Breechblock carrier and parts	
description and functioning	15
inspection	96
Brushes, care of	90
Buffer action	40-41

C

Cam, block rotating	17
Camouflage, paint as	92-93
Canvas webbing, cleaning contaminated	191
Care and preservation	
ammunition	163-164
brushes	90
carriage	86-89
cleaner and abrasives	89
cleaning after firing	79-80
fire-control equipment	
aiming circle M1	139
B.C. telescope M1915 or M1915A1	159
compass M2	148
one-meter base range finder M1916	154
prismatic compass M1918 (Sperry)	142

TM 9-331

155-MM HOWITZER M1 AND 155-MM HOWITZER CARRIAGE M1

C—Cont'd

	Page No.
Care and preservation—Cont'd	
general discussion of	78–79
howitzer	80–82
materials and tools, miscellaneous	90
organization spare parts and accessories	79
painting	91–93
preservatives	89–90
recoil mechanism	82–86
sighting equipment	117–119, 126
general discussion of	117–118
gunner's quadrant M1 or M1918	131
leather articles	118
lubricants	119
optical parts	118–119
telescope mount M25 with panoramic telescope M12	126
washing	90–91
Carriage, howitzer, 155-mm, M1	
care and preservation of	86–87
data on	7
description and functioning (See carriage under Description and functioning)	
electric brakes, care of	87–89
hand brakes	4
malfunctions	67–69
pneumatic tires and tubes	87
safety switch	4
serial number	95
storage and shipment	181–185
Casualties, lung irritant and vesicant	190
Cautions (See Precautions)	
Chamber capacity	4
Characteristics	
carriage	4
howitzer	2–4
trails	4
Charges, propelling	170–172
Chemicals, material affected by	
decontamination	189–191
general discussion of	189
protective measures	189
Classification of ammunition	160–161
Cleaners, list of	89
Cleaning	
after firing	
bore	79–80
breech mechanism	80
firing mechanism M1	80

	Page No.
ammunition exposed to vesicant gas	191
bore (and chamber)	58, 81
breech mechanism	75, 110
decontaminated material	190–191
firing mechanism M1	75, 102
primer holder	60
projectile	59
recoil slide	75
traversing and elevating racks and pinions	75–77
Cleaning and unloading rammer, M7	180
Cleanliness in lubrication	72–73
Cleats	182, 184
Clinometer	
dial	140–141
measuring angles of	147
Cold weather lubrication	73
Compass M2 (See also Prismatic compass M1918 (Sperry))	
accessories	144
adjustment	148
care and preservation	148
description	142–144
general discussion of	142
measuring angles (See Measuring)	
operation	145
orientation of grid	147–148
Compression, counterbalance spring	27
Control arc	17
Control rod stuffing boxes, oil drips from	68
Corrections (See Malfunctions and corrections)	
Counterbalance (mechanism)	
bracket	12
description and functioning	17–18
inspection	97
when to disassemble	110
Counterrecoil action	40
uneven or jerky	69
Counterrecoil and recuperator cylinder head box	35–36
Counterrecoil cylinder	
description and functioning	36
oil leaks from forward end	68
Counterrecoil rod, oil drips from	68
Cradle	32
(See also Recoil mechanism and cradle)	

196

INDEX

C—Cont'd

	Page No.
Crankshaft	
bushing	
description and functioning	16
inserting, illustration	111
description and functioning	15
removing, illustration	109
roller	16
Crating material for shipment	184
Crosshead	15
Current, checking in brakes	87–88
Cycles of breech mechanism functioning	24–27

D

Data	
carriage	7
howitzer	4–7
Decontamination of materiel	189–191
Description (See also Description and functioning)	
crankshaft	15
fire-control equipment	
aiming circle M1	135
B.C. telescope M1915 or M1915A1	156–157
compass M2	142–144
instrument light	137
one-meter base range finder M1916	148–149
prismatic compass M1918 (Sperry)	140
sighting equipment	
aiming post M1	126–127
gunner's quadrant M1 or M1918	128–131
telescope mount M25 with panoramic telescope M12	119–124
Description and functioning	
aiming post light M14	127–128
carriage	
axle and hubs	50–52
bottom carriage	44–46
electric brakes	53–54
elevating mechanism	42
equilibrators	43–44
firing jack	46–47
general	29–31
hand brakes	54
recoil mechanism and cradle	31–37
recoil mechanism functioning	37–41
top carriage	41–42
trails	48–50

	Page No.
traveling lock	47
traversing mechanism	42–43
wheels and tires	52
howitzer	
barrel assembly	9–12
breech mechanism	12–20, 24–27
firing mechanism	20–24, 27–28
general	9
Designation, howitzer	9
Detonation, projectile	59
Dimensions of carriage	7
Direct and indirect laying for elevation	124–125
Direct laying for azimuth	125
Disassembly	
block rotating roller	115
breech mechanism	103–111
firing mechanism M1	101–102
general discussion of	100–101
operating lever latch	115
Dragging brakes	85
Driver retaining ring, removing, illustration	108
Driver sleeve	16
Dummy projectile	166–168
unloading	62
Dust, operation in	187–188

E

Electric brakes	
care of	87–89
inspection	99
malfunctions and corrections	70–71
safety switch	53–54
Elevating, howitzer	58
Elevating mechanism	
description and functioning	42
inspection	98
Elevation, movement in	7
Equilibrators	
adjustment	44
alinement	43–44
description and functioning	43–44
Equipment (See Accessories)	
Exercising:	
recoil mechanism during storage	183
sighting and fire control instruments	187

Field report of ammunition accidents	178
Filling tube	180

TM 9-331

155-MM HOWITZER M1 AND 155-MM HOWITZER CARRIAGE M1

F—Cont'd

	Page No.
Fire-control equipment	
aiming circle M1	135–139
B.C. telescope M1915 or M1915A1	156–159
compasses	140–148
fuze setter M14	154
general description	135
graphical firing table M14	154–156
one-meter base range finder M1916	148–154
Firing howitzer	60–61
ammunition prepared for	165, 166
cleaning after	79–80
observation during	61–62
rate of	2–4, 7
Firing jack	46–47
handles, inspection	99
Firing mechanism M1	
assembly	102–103
breech will not open or fully close	66
cleaning and lubrication	75
description	9
description and functioning	22–24
adapter	21
general discussion of	20–21
housing	22
percussion mechanism	21–22
disassembly	101–102
function	27–28
inspection	97
removal	82
servicing	80
wrench	180
Firing pin	23
Firing position	
firing jack	47
howitzer	4
placing weapon in	55–56
Firing stresses	29
Firing tables	193
graphical, M14, M13 and M20	154–156
Fittings and oilers, lubrication	73
Floating piston, location	37
Freight car, securing howitzer and carriage on	181–184
Functioning (See also Description and functioning)	
breech mechanism	24–27
firing mechanism	27–28
Fuze	172–175
removing from shell	62

	Page No.
Fuze setter M14, description	154
Fuze wrench M7	180
Fuzing the projectile	59–60

G

	Page No.
Gas check pad, lubrication	82
Gas check seat, care of	81–82
Gas clouds (See Chemicals, material affected by)	
Gases, escape of	66
Grabbing brakes	71
Graphical firing table M14	154–156
Graphical firing tables M13 and M20	156
Gun Book, Artillery	179
Gunner's quadrant M1 or M1918	
care and preservation	131
description	128–131
general discussion of	128
operation	131
test and adjustment	131

H

	Page No.
Hand brakes	
carriage	4
description and functioning	54
Handspike	180
Hinge pin	17
Howitzer, 155-mm, M1	
alinement in mount	12
bore, lubricating	75
care and preservation	
breech mechanism	81
cleaning procedure	81
firing mechanism M1	82
gas check pad	82
gas check seat	81–82
general discussion of	80–81
inactivity	81
care during inactivity	81
data on	4–7
designation	9
description and functioning (See howitzer under Description and functioning)	
elevating	58
firing	60–61
inspection as a unit	96
loading	59–60
malfunctions and corrections	64–71
placing in:	
firing position	55–56
traveling position	62–63

INDEX

H—Cont'd

Howitzer, 155-mm, M1—Cont'd
	Page No.
returning to battery after firing	29
serial numbers	9, 95
storage and shipment	181–185
traversing	58
range	4
unloading	62
Hubs	50

I

	Page No.
Identification of:	
ammunition	161–163
lubrication points	72
Inactivity, care of weapon during	81
Indirect laying for azimuth	125
Inspection (See also Inspection and adjustment)	
arctic climates, materiel in	187
carriage as a unit	97–99
oil leakage in recoil mechanism	58
railroad cars	181–182
storage, materiel in	185
Inspection and adjustment	
carriage	97–99
howitzer	95–97
purpose	94
serial numbers	95
visual inspection upon receipt of materiel	94
Instrument light	137
Intermittent brakes	70
Interior surfaces, treatment of for storage or shipment	181
Intervals of lubrication	73

J

Jack float	46
"Jack knifing" of load, caution	63
Jerky counterrecoil	69

L

Leaks	67, 68
Leather	
care and preservation	118
cleaning contaminated	191
Liner of bottom carriage	44
Loading materiel for shipment	181–182
Loading rammer	180
Loading the howitzer	
closing the breech	60
projectile	
fuzing	59–60
preparing	59
propelling charge	60

	Page No.
Loading tray	180
Lost motion, testing sighting equipment for	125–126
Lot number, ammunition	161
Lubricants, supply of	73
Lubricating devices, painting	93
Lubrication	
arctic climates, materiel in	187
B.C. telescope M1915 or M1915A1	159
breech and firing mechanism	75
climatic conditions	73
gas check pad	82
general discussion of	72–73
recoil slide	75
reports and records	77
sighting equipment	119
traversing and elevating racks and pinions	75–77
Lubrication fittings, painting	72
Lunette	49–50
Lung irritant casualties	190

M

Magnet facing, care of	87
Malfunctions and corrections	
carriage	67–69
electric brakes	70–71
misfire	64–66
weapon	66
Maneuvering handles	49–50
Maneuvers, data on	7
Marking:	
ammunition	162
packings	176
breech ring	9
railroad cars	182
Materiel affected by chemicals (See Chemicals, materiel affected by)	
Measuring:	
angles of azimuth	
reading azimuth scale directly	147
reading reflected image	145
angles of clinometer	147
angles of site	147
azimuth	141
recoil, length of	61–62
Metal surfaces, painting	92
Misfire	
failures	
primer	64–65
propelling charge	65–66
precautions, general	64

TM 9-331

155-MM HOWITZER M1 AND 155-MM HOWITZER CARRIAGE M1

M—Cont'd

	Page No.
Model designation	161
Moist atmosphere, operation in	187
Muzzle, length of	4
Muzzle energy	7
Muzzle velocity	2, 4

N

Naphthalene, use as preservative	89–90
Nomenclature, standard, use of	160
Number, serial	9

O

Observation during firing	61–62
Obturator	
assembly of parts	13–14
description and functioning	13–15
Obturator spindle	96
Obturator spindle bearing sleeve, inserting, illustration	113
Oil index	
description and functioning	37
location	58
malfunctions and corrections	67–68
Oil leakage, inspecting for	58
Oil pump, M3	180
Oil release	180
Oil reserve	
gaging in recuperator	58
gaging in replenisher	57
Oilholes, painting	72
Oiling	
bore and chamber	80
breech mechanism	110–111
firing mechanism, M1	102
One-meter base range finder M1916	
care and preservation	154
description	148–149
general discussion of	148
operation	149–152
test and adjustment	152–154
Opening the breech	24–27
Operating lever	
description and functioning	17
does not latch properly	66
removing, illustration	109
Operating lever latch	
description and functioning	18
disassembly and assembly	115
Operation	
brakes	63
breech mechanism	55
elevating	58

	Page No.
fire control equipment	
aiming circle M1	137–139
B.C. telescope M1915 or M1915A1	157–159
compass M2	145
one-meter base range finder M1916	149–152
prismatic compass M1918 (Sperry)	140–141
firing	60–61
observation during	61–62
prior to	58
firing position, placing weapon in	55–56
general discussion of	55
loading	59–60
recoil mechanism, checking liquid in	57–58
shell, removing fuze from	62
sighting equipment	
gunner's quadrant M1 or M1918	131
telescope mount M25 and panoramic telescope M12	124–125
testing replenisher piston	84–85
traveling position, placing weapon in	62–63
traversing	58
under unusual conditions	
arctic climates	187
general discussion	186
moist or salty atmosphere	187
sandy or dusty conditions	187–188
tropical climates	186–187
unloading	62
Optical parts, care of	118–119
Organization spare parts	179
care of	79
(See also Accessories)	
Orientation on grid	147–148

P

Packing ammunition	176, 177–178
Paint(-ing)	
aiming posts	126
as camouflage	92–93
general discussion of	91
lubricating devices	93
metal surfaces	92
preparation for	91–92
projectiles	161–162
removing	93

INDEX

P—Cont'd	Page No.
Percussion hammer	
does not work freely	66
inspection	97
position	82
Percussion mechanism	21-22
Pintle, description and functioning	45-46
bearing	44-45
Piston, floating	37
Plunger, toothed rack of firing jack	46
Pneumatic tires and tubes, care of	87
Precautions in:	
disassembling breech mechanism	103-104
misfire	64
Preparing for firing	165
Preservation (See Care and preservation)	
Preservatives, naphthalene, flake	90
Preserving camouflage	92-93
Primer seat cleaning reamer	180
Primers	
description	175-176
failures	64-65
holder	
cleaning	60
description and functioning	24
Prismatic compass M1918 (Sperry)	
care and preservation	142
description	140
operation	140-141
test and adjustment	141-142
Projectile	
characteristics, table	167
cleaning	59
detonation	59
dummy	166-168
firing, preparation for	165, 166
fuzing	59-60
general discussion of	166
painting	161-162
preparing to load	59
service	166
travel of in barrel	7
weight of	7
weight zone markings	163
Propelling charges	
description	169-172
failures	65-66
preparing and loading	60
preparing for firing	165

Q	Page No.
Quadrant, gunner's, M1 or M1918 (See Gunner's quadrant M1 or M1918)	

R	
Ramps, use of for loading	182
Range indication, testing	153
Range movement of howitzer	4
Rate of fire, data on	7
Recoil, length of	4
Recoil, measuring length of	61-62
Recoil action	37-40
Recoil cylinder	
description and functioning	32-34
location	32
Recoil energy	29
Recoil fluid, application of	73
Recoil mechanism	
checking liquid	57-58
description	29
exercising during storage	185
functioning	
buffer action	40-41
counterrecoil action	40
general discussion of	37
recoil action	37-40
inspection	97-98
recoil oil	93
to establish recuperator oil reserve with oil pump	86
to establish replenisher oil reserve with oil pump	85
to exercise replenisher piston	86
to gage oil reserve in:	
recuperator	58
replenisher	57
to test operation of replenisher piston	84-85
Recoil mechanism and cradle	
counterrecoil and recuperator cylinder head box	35-36
counterrecoil cylinder	36
cradle, yoke, and cover	32
general discussion of	31
purpose	31
recoil cylinder	32-34
recuperator cylinder	36-37
replenisher	34-35
variable recoil mechanism	34
Recoil mechanism M6	
composition	31
on howitzer	4
serial number	95

TM 9-331

TM 9-331

155-MM HOWITZER M1 AND 155-MM HOWITZER CARRIAGE M1

R—Cont'd	Page No
Recoil oil, care of	83–84
Recoil rod, oil drips from	68
Recoil slide, cleaning and lubricating	75
Recoil system, air in	67
Records	
ammunition accidents	178
Artillery Gun Book	77, 179
Recuperator	
cylinder, description	36–37
gaging oil reserve in	58
oil reserve, establishing with oil pump	86
Regulator valve, location	37
Replenisher	
air in	67
description and functioning	34–35
gaging oil reserve in	57
oil leaks from rear of	67
oil, reserve, establishing with oil pump	85
piston	
exercising	86
stuck	67
testing operations of	84–85
Reports and records (See Records)	
Rifling, twist of	7
Rounds, components of	161
Rubber preservation	185
Rust prevention	78

S

Safety latch	18–20
Safety latch and plunger, removing, illustration	107
Safety switch	53–54
on carriage	4
Salty atmosphere, operation in	187
Sand, operation in	187–188
Securing howitzer and carriage on freight cars	181–184
Separate loading type ammunition	160
Serial numbers	95
howitzer	9
Service projectiles	166
Service rounds, unloading	62
Shell, removing fuze from	62
Shields	41–42
Shipment (See Storage and shipment)	
Sighting and fire control instruments, exercising in low temperatures	187
Sighting equipment	
adjustment	125–126

	Page No.
aiming post light M14	127–128
aiming post M1	126–127
bore sight	131–132
care and preservation (See sighting equipment under Care and preservation)	
general description of	117
gunner's quadrant M1 or M1918	128–131
preparation of for travel	125
procedure for bore sighting	132–134
telescope mount M25 with panoramic telescope M12	119–126
testing for lost motion	125–126
testing target	132
Socket wrench	180
Spades, location and description	50
Spare parts	
care of	79
general discussion of	179
Standard nomenclature used	160
Stop lights, connecting	88
Storage and shipment	
instructions, general	181
securing howitzer and carriage on freight cars	181–184
storage	184
water, shipment by	185
Strapping	182, 184
Subcaliber ammunition	176–178
Supports	182, 184

T

Telescope, B.C., M1915 or M1915A1 (See B.C. telescope M1915 or M1915A1)	
Telescope mount M25 with panoramic telescope M12	
care and preservation	126
description	119–124
operation	124–125
preparation for travel	125
test and adjustment	125–126
Test(-ing)	
aiming circle M1	139
B.C. telescope M1915 or M1915A1	158–159
gunner's quadrant M1 or M1918	131
one-meter base range finder M1916	152–154
prismatic compass M1918 (Sperry)	141–142

202

TM 9-331

INDEX

T—Cont'd	Page No.
Test(-ing)—Cont'd	
range indication	153
sighting equipment for lost motion	125-126
Testing target, use of	132
Tires	
description and functioning	52
storage treatment	185
(See also Wheels and tires)	
Tools	90
(See also Accessories)	
Toothed rack plunger of firing jack	46
Top carriage	
description and functioning	41
shields	41-42
trunnion bearings	41
inspection	98
Track, width of	7
Trail lock	49
Trails	
characteristics	4
description and functioning	48
lunette, battery box and maneuvering handles	49-50
spades	50
trail lock	49
Travel, preparation of sighting equipment for	125
Traveling lock	
description and functioning	47
inspection	99
Traveling position	
carriage	4
firing jack	47
placing howitzer in	62-63
Traverse, howitzer	58
range	4
Traversing and elevating racks and pinions, cleaning and lubricating	75-77
Traversing and elevating worm gear cases, servicing	75
Traversing handwheel, motion of	43
Traversing mechanism	
description and functioning	42-43
inspection	98

	Page No.
Tropical climates, operation in	186-187
Trunnion bearings	41
Tube, barrel assembly	9-12
Tube in mount, preventing rotation	9
Twist of rifling	7

U

Uneven counterrecoil	69
Unloading howitzer	62
Unpainted surfaces	
care of	87
treatment of for storage and shipment	181
Unusual conditions (See under unusual conditions under Operation)	

V

Valve, ball check	36
Variable recoil mechanism	34
Vent cleaning bit	179
Vesicant casualties	190
Visual inspection upon receipt of materiel	94

W

Washing ordnance materiel	90-91
Water, shipment by	185
Weak brakes	70-71
Weight zone markings	163
Weights, data on	4, 7
Wheel bearings, lubrication	75
Wheels	
carriage	4
description and functioning	52
Wheels and tires	
description	52
inspection	99
Wiring and controller, checking	87
Wrenches	180
care in using	100

Y

Yoke, description and functioning	32

203

TM 9-331

155-MM HOWITZER M1 AND 155-MM HOWITZER CARRIAGE M1

[A.G. 062.11 (3-13-43)]
[O.O. 461/36229 O.O. (3-18-43)]

By order of the Secretary of War:

G. C. MARSHALL,
Chief of Staff.

Official:
J. A. ULIO,
Major General,
The Adjutant General.

Distribution: D and H 7(2); IBn 6(2); Bn 9(2); IR 6(2); IC 6(14); C 9(2)

(For explanation of symbols, see FM 21-6)

Also Now Available!

Visit us at:

www.PeriscopeFilm.com

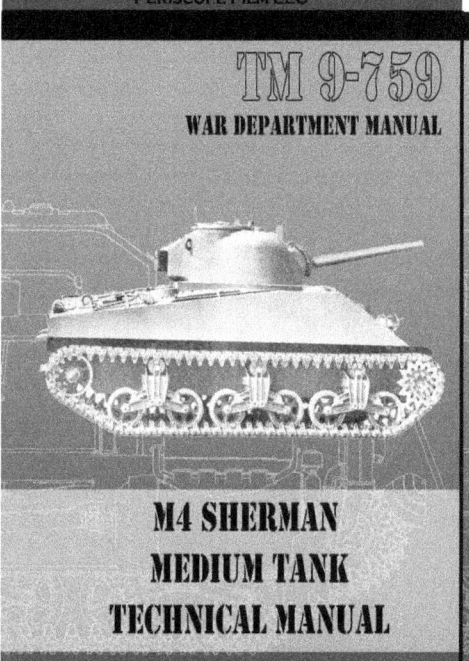

©2013 Periscope Film LLC
All Rights Reserved
ISBN#978-1-937684-37-2
www.PeriscopeFilm.com

www.ingramcontent.com/pod-product-compliance
Lightning Source LLC
Chambersburg PA
CBHW071704090426
42738CB00009B/1660